Code of Practice for Cyber Security in the Built Environment

Author: Hugh Boyes CEng FIET CISSP

The IET would like to acknowledge the help and support of CPNI in producing this document.

Published by The Institution of Engineering and Technology, London, United Kingdom

The Institution of Engineering and Technology is registered as a Charity in England & Wales (no. 211014) and Scotland (no. SC038698).

First published 2014

The Institution of Engineering and Technology
Michael Faraday House
Six Hills Way, Stevenage
Herts, SG1 2AY, United Kingdom

www.theiet.org

A list of organisations represented on this committee can be obtained on request to IET standards. This publication does not purport to include all the necessary provisions of a contract. Users are responsible for its correct application. Compliance with the contents of this document cannot confer immunity from legal obligations.

It is the constant aim of the IET to improve the quality of our products and services. We should be grateful if anyone finding an inaccuracy or ambiguity while using this document would inform the IET standards development team at IETStandardsStaff@theiet.org or The IET, Six Hills Way, Stevenage SG1 2AY, UK.

ISBN 978-1-84919-891-2 (paperback)
ISBN 978-1-84919-892-9 (electronic)

Contents

Participants in the Technical Committee

The IET would like to acknowledge the following organisations for their contributions to this document:

- Arup
- Atkins
- Centre for the Protection of National Infrastructure (CPNI)
- Corporate IT Forum
- Defence Science and Technology Laboratory (dstl)
- D'Arcy Race
- General Dynamics UK Ltd
- Imperial College London
- Martyn Thomas Associates Ltd
- Mike St John-Green Consulting Ltd
- PA Consulting
- Royal Institute of British Architects (RIBA)

Foreword

Attacks on buildings' operating systems are no longer hypothetical or simply the fictional narrative portrayed in films. In May 2010, Jesse McGraw ('Ghost Exodus') pleaded guilty to hacking a number of hospital computer systems in the United States, including the heating, ventilation and air conditioning (HVAC) computer, which was used to control the building service on two floors of the North Central Surgery Centre in Texas. In 2013 two security researchers successfully hacked into the building management system (BMS) used in Google's Sydney offices; a malicious hacker could have disrupted the use of the building or damaged building systems. The cyber-attack on the Target group of stores was initiated using remote access credentials from one of the company's HVAC contractors to gain access to the corporate network, resulting in the theft of card details for over 140 million credit cards.

If your answer to any of these questions is yes, you should carry on reading this section and decide who needs to take action.

(a) **Do you own an information asset that includes information about the design strategy, or construction of, your building and operation of its building systems?**

(b) **If this information asset was to be compromised, could this result in economic, operational, physical or reputational loss or damage?**

(c) **Do you own, operate or occupy a building, site or structure that has electronic or computer-based building systems?**

(d) **If the building systems were to fail, malfunction or were misused, could this result in economic, operational, physical or reputational loss or damage?**

Cyber security is not just about preventing hackers gaining access to systems and information. It also addresses the maintenance of the integrity and availability of information and systems, ensuring business continuity and the continuing utility of information assets. To achieve this, consideration needs to be given to protecting systems from physical attack, force majeure events, etc. and designing the building systems and supporting process to be resilient. Personnel security aspects are also very important, as the insider threat from staff or contractors who decide to behave in a malicious way cannot be ignored.

Increased training and awareness of health and safety have led to fewer serious accidents causing death, serious injury and damage to property. Failure to address cyber-security risks could lead to serious injury or fatality, disruption or damage to building systems, loss of use of the building, impact on business operations, financial penalties or litigation. Building owners, operators and occupiers need to understand cyber security and promote awareness of this subject to a building's stakeholders. This includes appropriate briefing of the design, construction and facilities management teams.

Our buildings are becoming increasingly complex and dependent on the extensive use of information and communications technologies at all phases of a building's lifecycle. This document explains why it is essential that cyber security is considered throughout a building's lifecycle and the potential financial, reputational and safety consequences that may arise if cyber-security threats are ignored. It provides clear practical guidance so that multi-disciplinary teams can understand how the management of various aspects of cyber security apply to their job roles and their personal responsibilities in maintaining the security of the building.

A building cyber-security strategy is relevant at every phase in the building's lifecycle, regardless of whether you are planning a new build, considering changes to an existing building, acquiring a building or operating an existing building. If the building does not have a cyber-security strategy and you answered yes to any of the questions at the start of this Foreword, you should create one now. This Code of Practice explains how to consider the cyber-security threats to the building, how to develop the cyber-security strategy and some of the practical issues relating to its implementation.

This Code of Practice uses principles rather than national legislation or specific standards to help promote good practice, and how it can be applied to managing cyber security in the built environment. The need for specific cyber-security measures to be taken will depend on the profile of the building, its use and the persons who occupy or use it. This Code of Practice focuses on building-related systems and, where they occur, all connections to the wider cyber environment. It provides a straightforward structure, centred on risk management, which supports detailed thinking about the activities and controls required to manage the cyber security of an organisation's building-related systems. It is important that the owners and operators of the building-related systems cooperate and collaborate with other users of information and communications technologies within the building, for example, those who operate the enterprise's desktop and central IT functions, and where the building contains control systems (such as industrial control systems) with the relevant operations and engineering departments.

This Code of Practice is intended to be used as an integral part of an organisation's overall risk management system and subsequent business planning, so as to ensure that the cyber security of building-related systems is managed cost effectively as part of mainstream business.

The cyber security of building systems is a relatively new field, which is growing in significance with the increasing use of information and communications technologies to deliver smarter buildings. This document is part of an ongoing programme of work led by the Institution of Engineering and Technology (IET). Previous publications include short briefing documents on intelligent buildings and the issues related to the impact of jamming and interference on wireless technologies. In 2013, the IET published a guide on the subject, *Resilience and Cyber Security of Technology in the Built Environment*[1]. This document builds on the concepts set out in the initial guidance, providing practical information on the steps that should be taken to implement appropriate cyber-security measures for building systems.

The IET is planning further work in this area including:

(a) periodic reviews and updates to this Code of Practice in the light of developments in industry;
(b) information on the ever changing landscape of threats and vulnerabilities that affect complex systems;

1 http://www.theiet.org/resources/standards/cyber-buildings.cfm

(c) guidance on the application of cyber security in the deployment of Building Information Modelling (BIM) across the full building lifecycle; and

(d) recommendations on the cyber-security skills required by those employed in the architecture, construction and engineering industries.

SECTION 1

Introduction

1.1 Aim and objectives

The aim of this Code of Practice is to provide guidance to a range of readers so that:

(a) the approach is systematic and reduces the residual risk to the cyber security of the building fabric and operations within the building;
(b) responsibilities and lines of accountability are clear and the right person does the right thing at the right time in the lifecycle of the building;
(c) cost effective security of the information and control systems can be achieved and the systems are secure by design; and
(d) features are worked into the building design from the outset so that costly rework is avoided.

To achieve this aim, the Code of Practice has the following objectives:

(a) to enable building owners and relevant stakeholders to create and implement an effective cyber-security management system for a building;
(b) to provide clear practical guidance so that multi-disciplinary teams can understand how the management of various aspects of cyber security apply to their job roles and their personal responsibilities in maintaining the security of the building; and
(c) to provide guidance that is easily understood and usable by a wide range of individuals from technical and non-technical backgrounds.

To achieve these aims, this Code of Practice identifies questions to ask and describes the issues to be considered; however, it is not intended to be a checklist of efficient cyber security for buildings. Unlike the cyber-security guidance published about generic IT or control systems, this Code of Practice addresses the complexity of both a building's and the stakeholders' lifecycles as the building progresses from concept through design, construction, operation and potentially eventual demolition.

1.2 Who should use this Code of Practice?

This Code of Practice is especially applicable to the Board or other functions that are required to take strategic decisions within an organisation. This Code of Practice is relevant to a wide range of job functions connected to the design, management, operation and security of any building-related systems, including those job functions responsible for:

(a) the financial and operational management of the building;
(b) the personnel and contractor security;
(c) ensuring that appropriate cyber-security policies and associated procedures exist;
(d) ensuring that appropriate procedures are implemented;

(e) specification and design of building systems, associated software and technologies;
(f) sale and systems integration of building-related systems;
(g) management of specific security tasks; and
(h) safe and secure operation of all equipment, plants and machinery.

It will also benefit individuals who wish to improve their knowledge of cyber security. It is applicable to a wide range of individuals who may not be from a traditional, trained-in-security background or experienced in managing building-related systems. As such they are not likely to have the specific knowledge associated with cyber security as part of their core competency.

1.3 Applicability

This Code of Practice is applicable to buildings that are associated with a wide range of organisations, irrespective of size, including the industrial, commercial and public sectors. The degree to which cyber security is a significant issue will vary from one building to the next, determined by the nature of the building, its use, its occupants, and the impact that a cyber-security incident could have on its use and the benefits that the building is designed to deliver.

In this document the terms 'building', 'building data', 'building systems' and 'building systems owner' are used. These terms are defined as follows:

(a) building – encompasses individual buildings, campuses, sites and structures, including their immediate physical environment.
(b) building data – encompasses any data, information, models and processes that are associated with the ownership, design and operation of a building. This includes the collaborative databases that are referred to as Building Information Modelling (BIM).
(c) building systems – those systems that are used to manage or control the cyber–physical systems in a building, which may include access control systems, building management systems, energy management systems and building fire, safety and security systems.
(d) building systems owner – the organisation that owns the systems and is ultimately accountable for the safe operation and maintenance of the systems.

1.4 Relationship with the building lifecycle

This Code of Practice can be applied to a building at any point in its lifecycle, including:

(a) during the specification, design, construction and operation of a building;
(b) when changes are made to the design and operation of an existing building;
(c) when there is a change of ownership of an existing building, a perceived risk or change in use.

The complexity of building and technology lifecycles and their interaction are examined in Section 2.

1.5 Document structure

The body of this Code of Practice contains the following sections:

(a) Section 2 – provides an overview of cyber security, typical cyber-security needs across a building's lifecycle and the role of key stakeholders.
(b) Section 3 – examines the application of cyber security across a building's lifecycle.
(c) Section 4 – outlines the next steps.

This Code of Practice is supported by a glossary of abbreviations and technical terms, and a number of detailed appendices providing more detailed information about cyber security, the creation of a cyber-security strategy, the development of cyber-security policy, processes and procedures, and the management of specific risk areas. The appendices are:

(a) Appendix A – Understanding cyber security
(b) Appendix B – Developing a cyber-security strategy
(c) Appendix C – Developing a cyber-security policy for a building
(d) Appendix D – Managing 'process and procedure' aspects
(e) Appendix E – Configuration control
(f) Appendix F – Managing 'people' aspects
(g) Appendix G – Managing technical aspects
(h) Appendix H – Trustworthy software
(i) Appendix I – Bibliography
(j) Appendix J – Factors to consider in assessing system context

SECTION 2

Overview

This section describes the generic building lifecycle used in this Code of Practice. It explains what cyber security is and identifies typical cyber-security needs across the building lifecycle. It also examines the building-related stakeholder roles that are potentially involved in maintaining the cyber security of a building.

2.1 The building lifecycle

Buildings typically have a lifecycle that can be measured in decades; they may undergo a number of changes during their lifecycle, including change of ownership, change of use, physical reconfiguration and upgrades or refurbishment of the structure, fabric, infrastructure, fixtures and fittings. For the purposes of this document, the building will be assessed from the perspective of the asset's owner and will be considered to have the generic lifecycle illustrated in Figure 2.1.

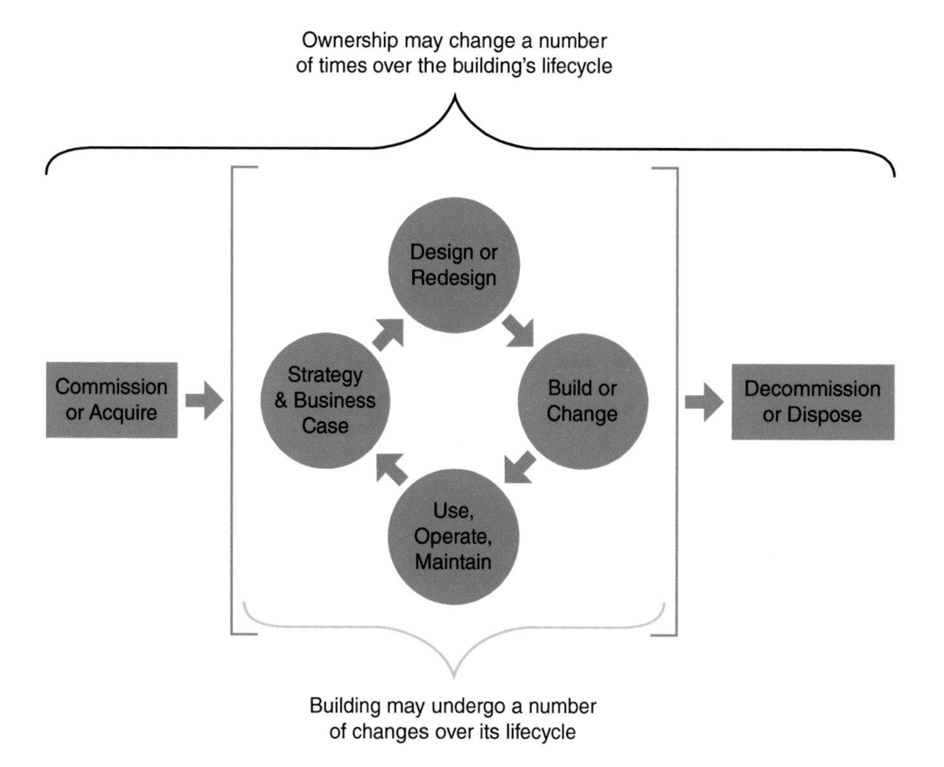

Figure 2.1 – A generic building lifecycle

The components of the lifecycle are:

(a) commission or acquire – from a building owner's perspective the lifecycle will start when they identify the need for a building and decide to commission the construction of a new building or acquire an existing one. From a tenant or building occupier's perspective the lifecycle may start when they 'acquire' the lease or the right to occupy the building.

(b) strategy and business case – the development of the rationale, requirements or project brief and the justifications for any change to the site occupied by the building, the building structure and services, its internal layout or use.

The formality, scope and format of the strategy, business case and project brief will depend on the organisation and the scale of the investment or change required.

(c) design or redesign – this covers the development of the project brief from the initial conceptual design of the building through to the detailed technical design. Depending on the complexity of the requirements and the scale of work required, the design may involve multiple phases or iterations.

(d) build or change – the implementation of the work required to deliver the project and hand it over for operational use.

(e) use, operate, maintain – the phase of the lifecycle of the building when it is being used in the way in which it was intended.

(f) decommission or dispose – from a building owner's perspective, the lifecycle will end when they decide to decommission the building or dispose of it. From a tenant or building occupier's perspective, the lifecycle will end when they 'dispose' of the lease or cease to occupy the building. Decommissioning may include the demolition of the building.

Figure 2.1 illustrates how the four core phases (detailed in b–e above) form a continuous lifecycle, reflecting the way that changes affect a building over its lifetime. These changes may be driven by a variety of factors, including change of building use, changes to the fabric and changes to building systems. This constant change introduces increasing cyber-security risks, through changes in personnel, responsibilities and building systems.

The relationship between the four core phases and a number of commonly used plans of work is illustrated in Figure 2.2.

Figure 2.2 – Relationship of lifecycle phases to commonly used plans of work

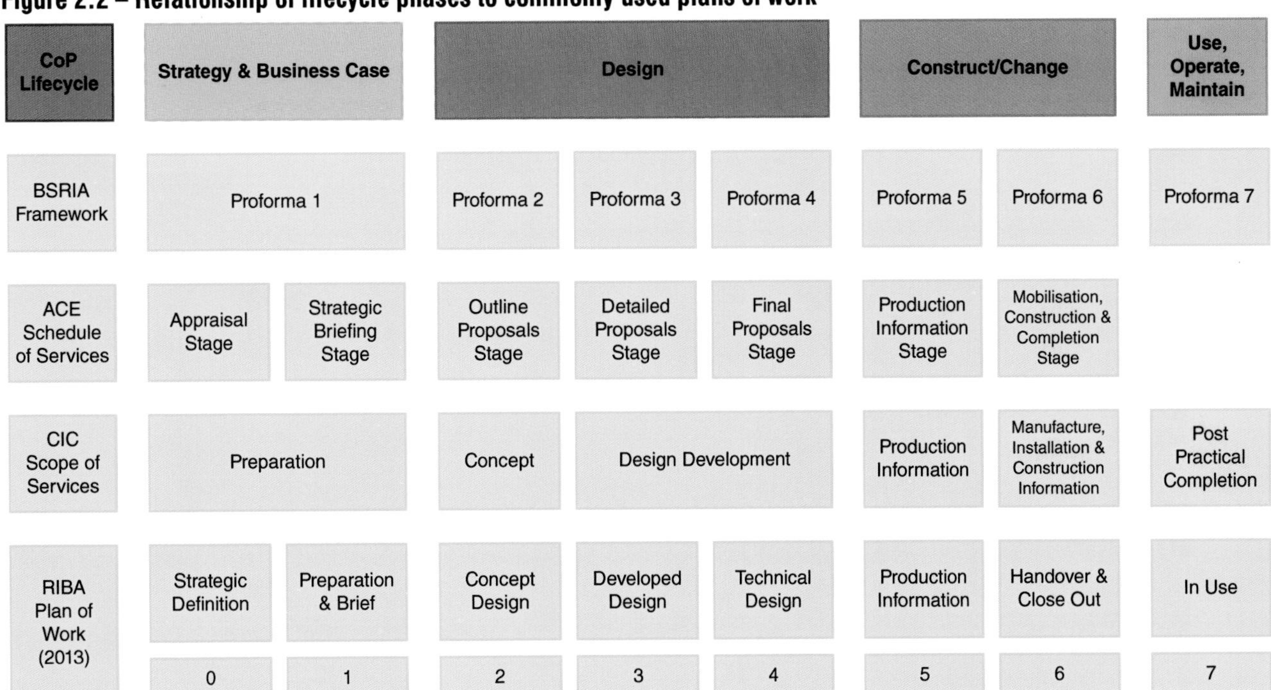

CoP Lifecycle	Strategy & Business Case		Design			Construct/Change		Use, Operate, Maintain
BSRIA Framework	Proforma 1		Proforma 2	Proforma 3	Proforma 4	Proforma 5	Proforma 6	Proforma 7
ACE Schedule of Services	Appraisal Stage	Strategic Briefing Stage	Outline Proposals Stage	Detailed Proposals Stage	Final Proposals Stage	Production Information Stage	Mobilisation, Construction & Completion Stage	
CIC Scope of Services	Preparation		Concept	Design Development		Production Information	Manufacture, Installation & Construction Information	Post Practical Completion
RIBA Plan of Work (2013)	Strategic Definition	Preparation & Brief	Concept Design	Developed Design	Technical Design	Production Information	Handover & Close Out	In Use
	0	1	2	3	4	5	6	7

From a cyber-security perspective, the need for a cyber-security strategy and its implementation could occur at any of the lifecycle phases, regardless of whether you are planning a new build, considering changes to an existing building, acquiring a building or operating an existing building. If the building does not have a cyber-security strategy, consideration should be given to creating one.

2.2 What is cyber security?

Cyber security is the collection of tools, policies, security concepts, security safeguards, guidelines, risk management approaches, actions, training, best practices, assurance and technologies that can be used to retain coherence and enable trustworthy operation of the organisation and users' assets in the cyber environment, where:

(a) the 'cyber environment' (also referred to as 'cyberspace') comprises the interconnected networks of electronic and computer-based systems, which therefore encompasses the internet (including an organisation's intranet), telecommunication networks, computer systems, embedded processors and controllers, and a wide range of sensors, storage and control devices. Given the interconnected nature of many systems, it cannot be limited to only those elements that are controlled or owned by the organisation; and

(b) the 'organisation and users' assets', which includes personnel, applications, services, social and business functions that exist only in cyberspace, and the totality of transmitted, processed and/or stored data and information in the cyber environment.

It is important to recognise that the cyber environment includes its critical supporting infrastructure, for example, utilities, uninterruptible power supplies (UPS), generators, etc. Experience shows that even standalone systems and isolated networks are at risk, from both attacks by malicious users and the introduction of malicious software via removable media. It should not be assumed that because a system is not apparently connected to the internet or any other network it is therefore secure. In practice there may be connections for ease of maintenance or system management, for example, to allow the download of patches and updates, which system users are not aware of.

Given the different lengths of lifecycles for the physical structure of buildings, their control systems, and typical enterprise IT systems and personal computing devices, this Code of Practice addresses cyber-security management through a set of seven attributes. By addressing these attributes, appropriate solutions may be adopted and adapted over the building lifecycle in response to changes in the building and technology, the use of the building and technology, and the nature and severity of potential threats. The seven cyber-security attributes, which are explained in Appendix A.3, are:

(a) confidentiality;
(b) possession and/or control;
(c) integrity;
(d) authenticity;
(e) availability;
(f) utility; and
(g) safety.

A more detailed explanation of cyber security is provided in Appendix A, which explains:

(a) cyber-security vulnerabilities, risks and threats;
(b) the seven cyber-security attributes relating to cyber–physical systems;
(c) how to establish the cyber-security context for a building by considering the following factors: people, their cyber-security awareness and understanding, the nature of building data, use of the electromagnetic spectrum, the building systems, infrastructure and wider environmental factors;

(d) types of threat agents that should be considered;
(e) available cyber-security standards, guidance and good practice.

2.3 Cyber-security needs by building lifecycle phase

The need for specific cyber-security measures will vary from building to building and across a building's lifecycle. This section considers some basic requirements that are likely to affect most non-domestic buildings[2]. The cyber security of the building depends on the creation and implementation of an appropriate security management regime that is based on:

(a) a regularly updated and reviewed cyber-security risk register;
(b) the cyber-security strategy for the building informed by the risks;
(c) the interpretation of the strategy into a cyber-security policy based on the building and building systems design;
(d) the design of appropriate cyber-security processes to implement the policy; and
(e) the use of repeatable cyber-security procedures based on the policy.

The relationship of this management regime to the building lifecycle is illustrated in Figure 2.3. To ensure a coherent approach to the management of cyber security across the lifecycle, if any of the elements from earlier phases in the lifecycle are missing these should be created as part of the current phase. For example, if the construction and the fit-out of a building is underway but there is no agreed cyber-security strategy or policy, one should be created to inform and guide the development of the processes and procedures that will be employed when the building is in use.

Figure 2.3 – Cyber-security management regime across the building lifecycle

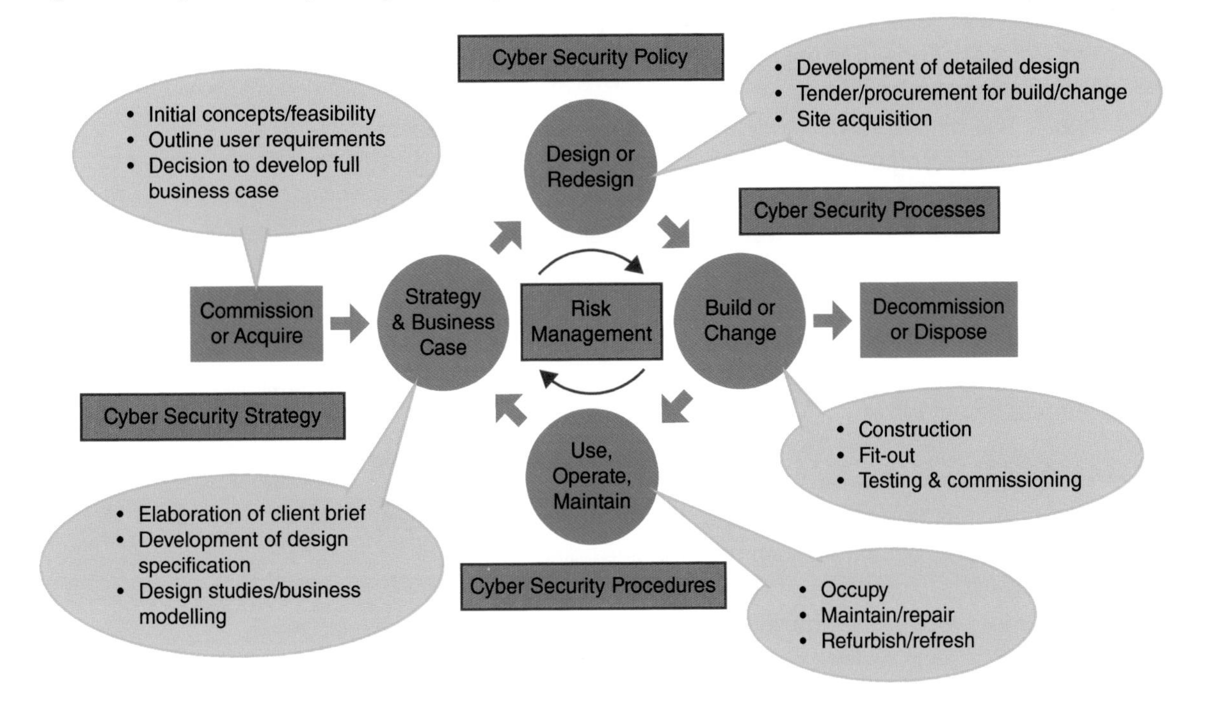

2 The risks in individual domestic buildings are currently relatively low, as few homes have high levels of automation. Increasing the use of 'smart' technologies within the home may change this in future. This Code of Practice is, however, applicable to large multi-occupancy residential buildings, for example blocks of flats, student residences, etc.

2.3.1 Commission or acquire – Phase A

During this phase the building owner or occupier will establish their need for a building. They will decide whether to build or acquire a building, either to become its owner or to lease it. Any decisions will relate to the strategic needs of the potential building owner or occupier. The key decision is often to undertake a more detailed study so as to develop a formal business case for the proposed action.

There will generally be a need for commercial confidentiality in this early phase of a major acquisition or building project. Failure to maintain good cyber security could result in the loss or premature disclosure of market-, price- or contract-sensitive information. In competitive situations where there is more than one party seeking to acquire the site or building, there may be the risk of commercial espionage where a competitor or third party seeks to acquire commercially sensitive information to undermine or compromise one or more of the bids[3]. Good cyber security needs to apply not only to the principals involved in the acquisition or commission, but to all professional advisers involved. Appropriate security advice should be sought from specialists, including the organisations' Chief Information Security Officer (CISO) or equivalent.

2.3.2 Strategy and business case – Phase B

Once an initial decision has been taken to commission or acquire a building, the next phase is the development of detailed business plans. A number of commercially sensitive activities may be undertaken, with sensitive information being created, exchanged and stored electronically, including:

(a) negotiations with financiers, investors and the existing site owner or their agent;
(b) preparation and submission of applications for planning consents;
(c) negotiation of options and or contracts related to the purchase, including obligations arising from planning applications or consents; and
(d) negotiations with existing occupiers of the site or building and with the owners and/or occupiers of neighbouring sites or buildings.

Whilst this phase has similar issues to the first phase, i.e. the need to protect commercially sensitive information, if the business plans involve significant changes to staffing levels or relocation there may be an increased risk of an insider threat. This may arise from a disaffected employee or contractor wishing to undermine the business planning.

As this phase includes the development of the project brief, there will be a need to address the cyber-security of any professional advisers involved. This is necessary to ensure adequate protection of:

(a) commercially sensitive data; and
(b) any sensitive building data about an existing operational facility, where harm or loss could occur following unauthorised disclosure.

Examples of the latter would include information about security features in an existing building, the location on floor plans of sensitive processes or the storage of hazardous materials, or the locations for storage of valuable and/or attractive items.

---

[3] See *The Economist and SANS* websites for information on industrial and corporate espionage respectively:
http://www.economist.com/blogs/democracyinamerica/2014/05/industrial-espionage
http://www.sans.org/reading-room/whitepapers/engineering/corporate-espionage-201-512

If the project relates to an existing building and it already has a cyber-security strategy, then during this phase the cyber-security strategy should be reviewed and updated. If the project relates to the commissioning of a new building or to an existing building that does not have a cyber-security strategy, then a strategy should be developed during this phase. Appendix B of this document explains how to consider the cyber-security threats to the building and how to develop the strategy.

A cyber-security strategy may require interventions that are specific to lifecycle phases, but there will also be interventions that can be made during any phase, for example, in response to a breach, when a new threat or change to an existing threat has been identified, or arising from advances in technology. Implementation of cyber-security countermeasures may occur at any point in the building's lifecycle; some will be an integral part of a construction-related project, others may arise on an ad-hoc basis either due to changes in response to, for example, new exploits or because of the need for additional protective measures following an incident.

2.3.3 Design or redesign – Phase C

As the project moves into the design phase the number of professional advisers involved will increase and, with the increasing use of BIM, the project brief and related design information may be widely available within the advisory team. Additional cyber-security measures may be required to protect certain aspects of the design (for example, physical security features, location and routes of alarm/CCTV cables, etc.), to reduce the risk to commercially sensitive information and to protect the design from unauthorised changes. The project owner needs to consider what cyber-security measures should be required of professional advisers and suppliers who have access to building data, and how the implementation of these requirements will be verified and/or enforced.

As part of the design process, the cyber-security policy for the building should be defined. Appendix C explains the need for this policy and how it relates to the cyber-security strategy, and to the more detailed cyber-security processes and procedures that will be produced during phases D and E respectively.

2.3.4 Build or change – Phase D

During this phase there will be three primary needs:

(a) to protect commercially sensitive information, including tender pricing information as the project team expands and additional personnel and professional advisers join the project team;

(b) to minimise the cyber-security risks to any data, information, systems or infrastructure in the building that are either used as part of the existing design or newly installed as part of the design; and

(c) to prevent damage or unplanned disruption to existing systems and infrastructure.

The protection of commercially sensitive information is a continuation of the processes and practices involved from the previous phases, but taking into account personnel and advisers that join and leave throughout this phase. Minimising the cyber-security risks to the building requires careful planning to avoid compromising existing systems or exposing new systems to a wide range of threats. The measures required will be determined by the detailed threat and vulnerability analysis undertaken as part of the cyber-security risk-management process.

During construction there may be a number of transient risks associated with the construction activities that may affect the availability or integrity of building systems, for example, the risk of a digger damaging buried power or communications cables. It is important to recognise that cyber security is not just about preventing unauthorised access to information (for further information, see Appendix A).

2.3.5 Use, operate, maintain – Phase E

This should be regarded as the steady state or normal operating state for a site or building. The protection of commercially sensitive information should be part of the business as usual function of operational departments and the enterprise's security and IT teams. From a building systems' perspective, this phase will require vigilance to ensure that cyber-security measures that have been put in place as part of the build (or change) phase are not compromised through ignorance, lack of maintenance, poor processes or lack of configuration and change control.

Cyber-security risks need to be frequently reviewed (for further information see Appendix D). As new vulnerabilities emerge and threats change, there is a need to monitor developments, assess the risk and update policies, processes and procedures where necessary or appropriate. There is also a need to implement cyber-security metrics to allow for an assessment of the effectiveness of existing procedures and countermeasures.

2.3.6 Decommission or dispose – Phase F

During the decommissioning or disposal phase of a building it is easy for mistakes to be made, which could lead to serious security incidents. Before a site or building, or a part of a site or building, is decommissioned and handed over for work to commence, or simply vacated, steps should be taken to ensure that cyber security is maintained, such as:

(a) the removal and secure storage or disposal of any special security equipment, for example, cryptographic systems;
(b) the removal and secure storage or disposal of any systems used to process or store personally identifiable or commercially sensitive data;
(c) the removal and secure storage or disposal of all other IT equipment in the area;
(d) the decommissioning of any public telecommunications network links or services;
(e) the decommissioning of any network or communications links to other companies' or organisations' sites;
(f) the decommissioning of any network or communications links to building management systems;
(g) the removal and secure storage or disposal of all used media, paper records, etc., containing personally identifiable or commercially sensitive data; and
(h) the re-routing of sensitive communications and network cabling routes.

The first time through this cycle (Phases B to E), or repeated iterations of it, generates a valuable asset for the physical state and related information about a building and its building systems that need cyber-security protection.

2.4 Stakeholder roles and cyber security

A stakeholder represents a specific category or group of individuals that have an interest in, or relationship with, the building or the built environment. There is a wide range of personas that could be involved at any phase in the building lifecycle, including:

(a) Owner/investor/landlord – should understand the risk that cyber-security threats can pose to the building and ensure that appropriate steps are taken during the design, build and use phases to protect the building, building systems and building data. Should ensure that there is a current assessment of the cyber-security risks to building systems and building data, and that appropriate steps are being taken to manage or mitigate any unacceptable building-related cyber-security risks.

(b) Occupier/user (lessee) – should liaise with the owner about the cyber-security context of the occupier's or user's operations. Should ensure that their own personnel, including contractors, suppliers, etc., comply with the cyber-security policy, procedures and processes for the building.

(c) Building system owner – has responsibility for ensuring that appropriate steps have been taken to consider the cyber-security risks to the building systems, that cyber-security policies, processes and procedures have been created, are in use and being maintained to protect the building, building systems and building data in accordance with an agreed cyber-security strategy for the building.

(d) Architect – has responsibility to the client for the building design and its implementation and should, in the light of the cyber-security risk assessment, take account of advice from specialist technical advisers on the cyber-security threats and suitable countermeasures.

(e) Consulting engineer – should advise the owner on cyber-security risks that affect the building, building systems and building data. Should ensure that other specialist technical advisers are aware of the cyber-security risks, and assist the owner with the development of the cyber-security policy for the building.

(f) Security adviser(s) – depending on the ownership and planned use of the building, one or more of these may be employed by the owner or specialist external advisers. The advice should cover all aspects of the building's security, i.e. personnel, physical and cyber security, as well as the security of the project team and supply chain.

(g) IT & communications (ICT) design consultant – should take account of the cyber-security risks and the building's cyber-security policy when working with the design team to develop an ICT solution for the building.

(h) Planner (consultants and local authority) – should take appropriate steps to protect any building data that is provided as part of the statutory planning and building regulations process.

(i) Other professional advisers – should take account of the cyber-security risks and the building's cyber-security policy when working with the design team to develop solutions for the building.

(j) Builder/prime contractor – should, whilst constructing and commissioning the building, ensure that agreed countermeasures are deployed to address identified cyber-security threats and should ensure that appropriate cyber hygiene is practiced by their staff, contractors and sub-contractors.

(k) Suppliers/sub-contractors – should comply with the cyber-security requirements as defined in their contracts, the building's cyber-security policy and any building- or site-specific cyber-security procedures.

(l) Systems integrators – should ensure that the integration of building systems is in accordance with the requirements of the building's cyber-security policy.

(m) Building services engineer – should design and implement and, where applicable, maintain the building systems in accordance with the requirements of the building's cyber-security policy.

(n) Facilities manager – should understand the structure and operation of all building systems, the cyber-security risks to those systems and the countermeasures in place or required to maintain the risks at an acceptable level. Should be responsible for security awareness and the training of personnel accessing the building systems and for managing connections from the systems to any third parties.

(o) Insurer – should understand the cyber-security risks to the building, the building systems and building data and how responsibilities are allocated for cyber-security matters.

(p) Lawyer – should ensure that, for the party that they advise, the contract or, where applicable, the lease, addresses cyber-security responsibilities and that appropriate obligations are in place to ensure an appropriate flow down within supply chain contracts.

How the involvement of stakeholders will vary across the building's lifecycle is illustrated in Table 2.1. The point at which individual stakeholders become involved will vary and this adds to the complexity of managing cyber security for the building and building data.

Table 2.1 – Stakeholder involvement across building lifecycle

Stakeholder involvement	**Lifecycle phase**					
	A	**B**	**C**	**D**	**E**	**F**
Owner investor/landlord	■	■	■	■	■	■
Occupier/user (lessee)	■	■	■	■	■	■
Building systems owner	■	■	■	■	■	■
Architect	■	■	■	■		
Consulting engineer		■	■	■		■
Security adviser(s)	■	■	■	■	■	■
IT & communications (ICT) design consultant			■	■		
Planner (consultants and local authority)		■	■			
Other professional advisers	■	■	■	■	■	■
Builder/prime contractor			■	■	■	■
Suppliers/sub-contractors				■	■	
Systems integrators			■	■		
Building services engineer			■	■	■	
Facilities manager	■	■	■	■	■	■
Insurer			■	■	■	
Lawyer	■	■		■		■

Lifecycle phases (see Section 2.3 for definitions):
A – Commission or acquire B – Strategy and business case
C – Design or redesign D – Build or change
E – Use, operate, maintain F – Decommission or dispose

Depending on the nature of the building, whether changes are being planned or are underway, there may be a number of stakeholder relationships to consider. The nature of these relationships will determine the allocation of cyber-security responsibilities between the parties involved. Examples of stakeholder relationships include:

(a) owner – occupier;
(b) owner – professional adviser;
(c) owner – building contractor;
(d) owner – facilities management contractor;
(e) owner – building systems owner;
(f) occupier/building systems' owner – occupier (for example, in multi-tenant buildings);
(g) occupier/building systems' owner – service provider/supplier; and
(h) occupier – individual, where the individual could be an employee, a visitor (authorised or unauthorised), a customer or a contractor.

The relationships illustrated above vary considerably in their formality and the degree to which the cyber-security responsibilities are formally allocated and accepted. For example, in the owner–occupier relationship the nature of any access by the occupier to building systems, associated processes and building data is likely to be covered in some form of contract, tenancy agreement or lease. In this situation it is possible to impose obligations on both parties, for example, in respect of data protection, system access and acceptable use. The relationship with the supplier can be more complicated and, whilst the contract may set out responsibilities, it is the behaviour of individual employees of the supplier that will determine the effectiveness of the allocation of cyber-security responsibilities. For example, careless use of removable media by a third party's employee could easily result in malware infections.

Three specific situations will prompt necessary measures for the protection of building systems:

(a) when a third party has remote access or connections to the building systems, associated processes and building data;
(b) when a third party has on-site access to the building systems, associated processes and building data, either through IT equipment already installed in the building or through their own devices (laptop, smartphone, etc.); and
(c) when an insider uses authorised or unauthorised routes to affect any of the cyber-security attributes listed in Section 2.2.

SECTION 3

Applying cyber security through the lifecycle of a building

3.1 Introduction

As illustrated in Section 2.1, buildings have complex lifecycles and typically undergo a number of change and technology refresh cycles. The use of the building may also change significantly over its lifecycle, as may the potential vulnerabilities and the threat landscape. Rapid changes in technology or business requirements can lead to changes to building systems and operations that were not or could not be foreseen at the time it was designed, last refurbished or refitted.

An example of an operational change that introduces a new cyber-security risk would be if a new facilities management contractor were to propose significant cost savings by implementing remote monitoring of building systems, plant and machinery. The potential presence of such remote connectivity may not have been considered during the original building systems design and appropriate countermeasures may not have been implemented to prevent or limit damage from attacks mounted via this remote connectivity.

Cyber security impinges both on new buildings that are still at a concept or design stage and existing buildings that are being acquired or changed. This section considers the potential impact that poor cyber security can have on buildings and their stakeholders.

3.2 Who is responsible for the cyber security of building systems and data?

Responsibility for the cyber security of a building, its systems, associated business processes and the building data should be addressed at both strategic and operational levels.

At a strategic level:

(a) the building owner is responsible for ensuring there is a cyber-security strategy covering the building systems and building data; and

(b) the building owner and occupier(s) are jointly responsible for the building's cyber-security policy. This joint responsibility is necessary to ensure that the policy covers aspects that are under the control of both parties and that the policy has been translated into processes and procedures to ensure its implementation.

At an operational level, in collaboration with the building owner and occupier(s):

(a) the building systems' owner is responsible for ensuring the processes and procedures are used on a day-to-day basis; and

(b) the building systems' owner is responsible for maintaining the processes and procedures to reflect changes in the business systems and associated processes, and in response to emerging threats.

3.3 What building systems and data need to be protected?

The need to protect any of the above types of asset or item should be determined on a case-by-case basis. The building owner and relevant stakeholders should consider what value they place on an asset and its associated benefits to determine the potential impact of specific cyber-security risks.

Depending on the building, its use and its stakeholder needs, the following types of asset or item may need to be protected:

(a) information or data related to:
   i. an owner's or investor's business plans or strategy for a site or building;
   ii. the design and technical operation of the building;
   iii. contractual data related to the design, construction and operation of the building and any changes to the building;
   iv. the use of the building and business operations within it; and
   v. personally identifiable information (PII) about building occupants or users.

(b) building systems, in order to:
   i. prevent unauthorised changes to, or use, control and operation of, technical systems, including:
      - HVAC;
      - building management systems;
      - access control systems;
      - CCTV and alarm systems; and
      - industrial and process control systems;
   ii. prevent loss of commercially or operationally sensitive information or data from building-related systems; and
   iii. prevent loss of information affecting an individual's privacy or security, for example, images from CCTV systems or PII held in building systems;

(c) operational information related to the use of the building or its occupiers, for example, where sensitive operations are housed, or where valuable items or dangerous materials are stored.

3.4 What could adversely affect the building systems and data?

The increasing reliance of organisations on information and communications technologies means that interference with, or the failure of, electronic and computer building-based systems can have a serious impact. The cyber-security attributes related to things that can go wrong and that could adversely affect the assets or items are outlined in Section 2.2 and Appendix A.3. The level of protection required will generally be determined by the potential impact on building owners and stakeholders compared to the cost of implementing appropriate countermeasures. Examples of incidents that could affect these attributes include:

(a) confidentiality – the unauthorised disclosure of detailed business plans related to the acquisition and redevelopment of a site by a disgruntled employee that has a substantial impact on commercial negotiations regarding its acquisition and use.

(b) possession or control – the infection with encryption malware of computers used by a design team could result in critical design information being held to ransom; it is still in the possession of the team but they have lost control of it and cannot use it.

(c) integrity – the control of lighting and ventilation in individual conference rooms is managed using wireless controllers. Their operation becomes intermittent and, on investigation, there is an electromagnetic interference problem caused by a local radio frequency source interfering with and degrading the control signals.

(d) authenticity – a purchase of heavily discounted software, which turns out to be counterfeit with malware included on the source disks, is installed on the users' enterprise network and has spread to a number of clients, causing both reputational and financial damage.

(e) availability – the use of cloud-based energy management services is badly disrupted due to a denial of service attack on the cloud service provider, leading to energy savings losses and intermittent power outages.

(f) utility – during installation and commissioning of new site-wide access control system, the file and data conversion process employed for the transfer of access control permissions results in corruption of critical data. As a result the data has to be manually recreated and entered into a new system, disrupting site operations and incurring significant extra costs.

(g) safety – hacking of remote support connection on fire alarm systems results in fire alarms being set off on three consecutive days, with response by fire brigade on each occasion. Fire certificate is withdrawn by fire authorities rendering building unusable on safety grounds until the fire alarm system is modified to prevent reoccurrence of the security breaches.

The above examples illustrate that cyber security is not just about addressing the potential threats from malicious threat agents, for example hackers. Threats can emanate from defects in system or process design, or from human negligence or error.

3.5 Where are the building systems and data located?

The physical location of building systems is a significant factor in the exposure of the systems, their components and infrastructure to a variety of threats. For example, unauthorised physical access to a workstation can allow the system to be compromised through use, intentional or otherwise, of removable storage media infected by malware (for example, USB storage devices, CDs, DVDs, etc.). The location can also determine the susceptibility of system components to physical damage or interference, by accident or design, or by natural causes. Depending on the nature of the building and the planned operation and maintenance arrangements for the building systems, it may be appropriate to consider physical collocation of the processing elements of these systems with other central IT systems in the building, for example, in the building's computer rooms.

The systems used to create, process, manage, store or display the building data may be located:

(a) within the building or site;
(b) in other accommodation used by the owner, operator or occupier; or
(c) with a third party service provider, for example, the facilities manager, cloud service provider, etc.

By understanding the location of a system's components and their criticality to the correct and continuing operation of the system, the risks can be assessed and appropriate controls or countermeasures adopted. Appendices G.1 and G.2 provide further information on operational and physical security considerations.

Identification of the location of the building systems enables protection of the building data that is generated, stored and processed on these systems, but that will not address all building data. For example, it is unlikely to cover some strategic information assets related to the design and operation of the building, BIM related information created as part of a design process, etc. Consequently, in addition to considering the location of building systems, the location of building data throughout the building's lifecycle is taken into account. This enables the risks to this information to be assessed and appropriate controls or countermeasures adopted.

The building data used by the building systems or required for the design, operation and management of the building may be stored on site in the building or in an adjacent building, or it may be held off site:

(a) by the building owner, operator or occupier, or their advisers or representatives;
(b) by a building-related professional or supplier, for example, plans, detailed designs, configuration and maintenance information;
(c) by regulatory or statutory authorities, for example, planning applications and building regulations information;
(d) by suppliers or service providers; or
(e) by third parties, for archive, back-up or business continuity purposes.

The location of building data is likely to change throughout the building lifecycle. This should be addressed as part of the operational management of the data (see Appendix G.1 for further information on operational security). The provision of a suitable environment for the management and security of the BIM related information needs to be considered throughout the building lifecycle, which may be complicated by changes in building ownership, occupation and operation.

3.6 How should building systems and data be protected?

Our aim in managing cyber security is to retain coherence and enable trustworthy operation of the organisation's and user's assets in the cyber environment. The measures required to achieve this fall into one of four categories:

(a) physical – to protect the building systems and data from damage, theft or interference;
(b) technical – the electromagnetic, physical and logical design and implementation to enable and achieve trustworthy operation of building systems;
(c) procedural – the processes and procedures relating to the use of building systems and data throughout their lifecycle;

(d) personnel – the background checks, vetting, training and education of individuals who will have access to, or use of, the building systems and data.

The choice of measures will depend on the building's cyber-security context and the level of risk associated with perceived threats and vulnerabilities. In assessing the measures to be applied, consideration should be given as to the degree of alignment that may be practically achieved between those in place for enterprise systems and those applied to building systems. More detailed information on protective measures in the above four categories can be found in the appendices to this document and the references in the bibliography.

Next steps

If you own, operate or occupy a building that has electronic- or computer-based building management systems and associated information assets, what would the consequences be if the systems were to fail, malfunction or be misused? If the answer to this question is that the result could be economic, operational, physical or reputational loss or damage, then steps should be taken to address the cyber security of the building systems.

A starting point for this is to establish whether these systems have a cyber-security strategy. If they do not, then steps should be taken by the building owner and/or building occupier to create one. Appendix B provides guidance on how to develop a cyber-security strategy for a building.

Once there is a cyber-security strategy in place for the building, there will be a need for a building cyber-security policy and the supporting processes and procedures. The development of these management products is described in Appendix C. When developing the policy, processes and procedures, a number of specific areas will need to be addressed:

(a) Appendix D explains areas that need to be addressed from a management perspective, including the identification and handling of risk, supply chain security, systems operations, and incident handling;
(b) Appendix E explains the need for configuration control to manage systems architectures, inventories of hardware and software, and the configuration of systems;
(c) Appendix F explains issues related to people aspects, including the need for cyber-security awareness, education and training;
(d) Appendix G explains various technical aspects related to the cyber security of building systems; and
(e) Appendix H explains the need for trustworthy software and good practice related to software engineering across the full software lifecycle.

Where the organisation that owns and operates the building systems has a CISO and suitably skilled cyber-security team that understand the issues associated with configuring, managing and operating complex control systems, they may be able to support the building engineering and facilities management teams in the design and operation of building systems. However, it is important to recognise that some of the tools and techniques that may be used in an enterprise IT environment could be wholly inappropriate when applied to building systems. This Code of Practice recommends that where appropriate and applicable an enterprise's corporate IT standards and its cyber-security training and awareness material may be used to inform the safe and secure design and operation of building systems and related business processes. Where there is any doubt about suitability of the corporate standards, specialist advice should be sought.

For further guidance on cyber-security standards and their implementation, a number of documents are listed in Appendix I. To assist the development of the cyber-security strategy, policy, process and procedures, a set of questions are provided in Appendix J. These questions are not intended to be a checklist, but instead aim to probe the nature, operation and configuration of building systems and data.

This document sets out a comprehensive approach to the management of cyber security as it affects a building or other structure and their associated building data in the built environment. The approach adopted and the requirement for specific measures will vary from building to building, and may also vary over the life of a building, in the light of changes of use, occupier and variations in the perceived threat environment. In the case of multi-tenanted buildings there may be a need to review existing policies, processes and procedures on a change of tenant, and individual tenants may have differing security needs. The approach advocated in this document can be tailored to a building and its stakeholders' needs.

The need for any particular cyber-security measures should be determined following a risk assessment and taking account of the overall cyber-security strategy and policy for the building, its building systems and data. The assessment should be conducted in a pragmatic, appropriate and cost effective manner, taking into account the nature of the building and the impact of failures and/or disruption experienced by the stakeholders.

The risk and cyber-security situation will never be static, so the risk assessment cannot simply be handled as a 'one-off exercise'. There will be a need to review and update the analysis periodically, based on a combination of elapsed time since the last assessment, changes to the nature or use of the building, changes in technology and the potential threat to the building, the building systems and building data. The frequency of reviews and potential triggers for an ad-hoc review should be considered when establishing the cyber-security strategy for the building.

If the building you are planning, constructing or occupying does not have a cyber-security strategy and the cyber-security risks have not been assessed, now is a good time to start. Don't wait until an incident has occurred; take the first step towards reducing any cyber-security risks to the building by conducting a risk analysis and defining the building's cyber-security strategy.

GLOSSARY

Abbreviation/Term	Meaning
2FA	Two factor authentication, or 2FA, is a two-step verification process used as part of an access control system, for example when accessing a building or logging into an online account. Examples of 2FA include using a smartcard and personal identification number (PIN) or a username/password and a further piece of secret information or token generated PIN.
Administrative Privileges	The highest level of permissions that is granted to a computer user typically granted to systems administrators.
AFD	Anticipatory failure determination – process by which a user can thoroughly analyse a system for failure mechanisms in order to obtain an exhaustive set of potential failure scenarios and identify solutions to prevent, counteract, or minimise the impact of these failure scenarios.
APT	Advanced persistent threat – a form of cyber-security attack, often used for intelligence gathering, involving an attacker who has the capability and intent to persistently and effectively target a specific individual or organisation.
Asset	This may be a physical or logical item that has some inherent benefit, value or use for its owner or user. Physical items will include the building and building systems. Logical items include data, information, software or digital content. Intellectual property is an asset, whether stored in physical or logical form.
BIM	Building information modelling – digital representation of physical and functional characteristics of a facility creating a shared knowledge resource for information about it, which will become a reliable basis for decisions during its lifecycle, from earliest conception to demolition.
BIS	Department for Business, Innovation and Skills
Bluejacking	A hacking method that allows an individual to send anonymous messages to Bluetooth-enabled devices within a certain radius.
Bluesnarfing	A hacking method performed when a wireless, Bluetooth-enabled device is in discoverable mode. It allows unauthorised remote access to data on the target Bluetooth device.
Bluetooth	A wireless technology standard [IEEE 802.15.1] used for communicating data over short distances and that may be used to create personal area networks.

Abbreviation/Term	Meaning
BMS	Building management system
BYOD	Bring your own device, refers to the use of an employee's personal computing device (for example, smartphone or tablet computer) to access and work on their employer's systems, for example to access emails or other corporate/enterprise applications.
CERT	Computer emergency response team
CESG	Information Security arm of the Government Communications Headquarters (GCHQ)
CISO	Chief information security officer
COMSEC	Communications security – the practice of preventing interception or access to communications traffic.
Command and control server	A command and control server (C&C server) is a computer that is used to issue commands to and receive data from compromised devices as part of a malware distribution or hacking network.
Continuity	The unbroken and consistent operation of a system, process or business over a period of time.
Convergence	The tendency for previously separate technologies, for example, voice, data, video to now share resources, both physical (such as cabling) and logical (processing, storage, etc.), and to interact with each other.
COTS software	Commercial-off-the-shelf software
Countermeasures	A measure introduced to reduce either (a) the likelihood that a vulnerability will be exploited, or (b) the impact in the event that a vulnerability is exploited.
CPD	Continuing professional development
CPNI	Centre for Protection of National Infrastructure
CRM	Customer relationship management
Cyber–physical system	A cyber–physical system is broadly defined as integration of computation, networking and physical processes. Examples include control systems, where there are sensors and actuators allowing interaction between the physical and cyber domains.
Defence-in-depth	The coordinated use of multiple (layered) security countermeasures to protect computers and networks. Based on the military principle that a complex, multi-layered defence system is more difficult to penetrate than a single barrier.

Abbreviation/Term	Meaning
DoS	Denial of Service – an attack on a network or system that is designed to bring prevent the normal use of the system. May be achieved by flooding networks with traffic, jamming wireless communications or overloading a processing system with requests.
DDoS	Distributed denial of service – an attack by multiple systems that tries to flood a targeted system with traffic until connection requests are refused and the targeted system fails to respond to existing connections.
DMZ	De-militarised zone – a physical or logical sub-network protected by firewalls used to share data between trusted and untrusted networks, for example, between an organisation's intranet and the internet.
DNS	Domain Name System (or Service or Server), the internet service used to translate domain names (for example, www.theiet.org) into individual IP addresses.
EACOE	Enterprise Architecture Center of Excellence
EMC	Electromagnetic compatibility
EMI	Electromagnetic interference
ERP	Enterprise resource planning
FMEA	Failure mode and effects analysis – a systematic approach for the identification of possible failures in a design, process, product or service. 'Failure modes' are the ways, or modes, in which something may fail.
FTP	File transfer protocol, an internet protocol used for transferring files between computers.
GNSS	Global navigation satellite system, a generic term for a satellite navigation system with global coverage.
Governance	The management, decision-making and leadership processes employed by an organisation to ensure consistent and cohesive management of a given area of responsibility.
GPS	Global positioning system, generally used to refer to the United States' Global Positioning System (GPS), which is currently the world's most utilized satellite navigation system.
GSM	GSM (Global System for Mobile communications) is an open, digital cellular technology used for transmitting mobile voice and data services.
GSM modem	A modem designed for transmitting digital data over a GSM network.
HMI	Human Machine Interface, typically provided in computer installations by screens, keyboards and pointing devices (for example a mouse, touchpad or tracker ball).

Abbreviation/Term	Meaning
HoMER	Holistic Management of Employee Risk, a personnel management framework developed by CPNI, see Appendix F.1.3.
HVAC	Heating ventilation and air conditioning
ICS	Industrial control systems
ICT	Information and communications technology
IDS	Intrusion detection system
iOS	The operating system used on Apple Inc.'s mobile devices, for example, iPhones and iPads.
IP address	The Internet Protocol (IP) is a means of identifying a device on a TCP/IP network. It comprises a unique string of numbers separated by full stops used to route network communications.
IPS	Intrusion prevention system
Insider risk	The risk arising from the behaviour or actions of an insider. An insider is someone who exploits, or has the intention to exploit, their legitimate access to assets for unauthorised purposes.
LAN	Local area network
LTE	LTE (Long Term Evolution) is a wireless broadband communications technology that will enable users to transmit data at approximately 10 times faster than the current 3G GSM network.
LTE modem	A high-speed wireless broadband modem used to transmit data over an LTE network.
Malware	Malicious software used to attack, disrupt, compromise security, or take control of a computer system or individual computer.
MRP	Materials resource planning
NFC	Near field communications
Password complexity	A partial measure of the strength or effectiveness of a password in resisting others from guessing the password and brute-force attacks. A complex password will include a mixture of numbers, letters and special characters arranged in an unpredictable order. The overall strength is a function of complexity and password length.
Phishing	The practice of sending fraudulent emails purporting to be from a reputable organisation or person so as to induce recipients to reveal personal or confidential information online.
PII	Personally identifiable information
PLC	Programmable logic controller
Reconnaissance	The preliminary inspection, survey, exploration and/or research undertaken when contemplating an attack or specific action.

Abbreviation/Term	Meaning
Resilience	The ability to withstand a level of failure or disruption and to adapt or respond to dynamic internal or external changes whilst continuing to operate with limited impact on the organisation or business.
RFC	May be used in a configuration or change management system to refer to a request for change. Can also refer to a Request for Comments (RFC), a formal document from the Internet Engineering Task Force (IETF) that is the result of committee drafting and subsequent review by interested parties. These documents are used to develop and document technical standards for the internet.
Risk	The probability or threat of damage, injury, liability, loss, or any other negative occurrence that may be caused by vulnerabilities, and that may be avoided through pre-emptive action, such as the implementation of appropriate countermeasures.
RF	Radio frequency
RFID	Radio frequency identification
RTU	Remote terminal unit – part of an industrial control system
SCADA	Supervision control and data acquisition
Security token	A device used as part of a security authentication process, typically associated with logging in to a system or securing communications.
SLC	Security level capability
SLT	Security level target
SMS	Short message service – the text service used for communication between phones/mobile phones.
Spear phishing	A variant of phishing, but the email is from a person you know or an organisation you have a connection to.
Social engineering	The practice of obtaining confidential or sensitive information by manipulating legitimate users. A social engineer will commonly use the telephone or internet (for example, email or instant messaging) to trick a person into revealing the information or getting them to do something that is against typical operational policies.
SRD	Short range devices
SSL	Secure Sockets Layer (SSL) is a standard security technology that is used to establish an encrypted link between two computers, typically a server and a client. SSL may be used to protect web browsing, access to an email client, etc. from interception as the communication passes over the network.
STEEPLED	An abbreviation for a number of environmental factors: Societal, Technological, Economic, Environmental, Political, Legal, Ethical and Demographic. Used in environmental scanning.

Abbreviation/Term	Meaning
Structured cabling	The implementation of a structured building or campus telecommunications (i.e. telephony, computer network, video, etc.) cabling infrastructure that comprises a number of standardised smaller elements.
Systems integration	The process of bringing together component systems or sub-systems to create a single system whose functionality operates as a coordinated whole.
TCP/IP	Transmission Control Protocol/Internet Protocol (TCP/IP) is the suite of communications protocols that are used to connect devices on the Internet.
TCP/IP port	The port refers to a number assigned to user sessions and server applications within an IP network. Port numbers, which are standardized by the Internet Assigned Numbers Authority (IANA), reside in the header area of the packet being transmitted and thus identify the purpose of the packet (for example web, e-mail, voice call, video call, etc.).
Threat agent	An individual or group that can manifest a threat, this includes individuals, groups and natural effects.
Threat	Anything that is capable of acting in a manner resulting in harm or hazard to an asset, individual, organization or the environment.
TL	Trustworthiness Level
TSF	Trustworthy Software Framework
TSI	Trustworthy Software Initiative, a public good initiative funded by the UK Government to improve the quality of software. The Initiative has produced BS PAS 754 and the Trustworthy Software Framework.
UPS	Uninterruptible power supply – an emergency power system providing continuity of supply in the event of an interruption in the mains power supply.
VLAN	Virtual local area network – configuration of a network to create broadcast domains that are mutually isolated.
VPN	Virtual Private Network, a private network configured within a public network such as the internet. Data encryption is used to maintain confidentiality and privacy. VPNs allow mobile or remote users access to the company LAN without being physically connected to the network, for example when working off-site.
Vulnerability	A weakness or susceptibility of an asset or group of assets that can be exploited by one or more threats. The vulnerability may arise from an aspect, flaw or weakness in the design, implementation, or operation and management of a system or process that could be exploited.
WARP	Warning and reporting point, a centre used by organisations collaborating over cyber-security alerts and incidents.

Abbreviation/Term	Meaning
WEP	Wired equivalent privacy – a security algorithm for IEEE 802.11 wireless networks. Has been superseded by Wi-Fi Protected Access (WPA).
Wi-Fi	Technology used to deliver a wireless local network based on the IEEE 802.11 standards.
Wireless	Used in this document to cover networks and communications carried by radio frequency transmissions allowing the passage of information or data without a physical (wired) connection.
WPA	Wi-Fi Protected Access
ZigBee	A specification for a communications protocol using small, low power digital radios, based on the IEEE 802.11 standard for personal area networks.

APPENDIX A

Understanding cyber security

A.1 Introduction

This appendix describes an approach to managing cyber security based on clear practical principles and an analysis based on seven key attributes. Applying these principles and the analysis in a systematic rather than in an ad-hoc or 'fire-fighting' fashion allows an organisation to adopt a holistic approach to cyber security. This in turn reduces the need for expensive and/or disruptive gap-filling activities when a cyber-security incident arises, to which an organisation needs to react. As a result, personnel in the organisation and, where applicable, supporting contractors and suppliers will have a better understanding of particular cyber-security issues and how to cost effectively mitigate them.

The varied nature of cyber-security threats and incidents suggests that there is no single strategy that is capable of addressing all cyber-security risks. The rate of change of technology and the steady flow of serious vulnerabilities in operating systems, software libraries and applications, together with ever changing threat agents, means that any strategy needs to be kept under regular review. Business change also has a significant impact on cyber security, for example, the introduction of bring-your-own-device (BYOD) into a building, particularly the building services area, can have a significant impact on the risk profile. There is also a trend to deliver some assets as services, for example, the provision of back-up or standby power supplies under the management and control of a third party.

A multi-faceted approach is therefore needed to manage the cyber-security risks in the face of the changing cyber-security environment. This should include a proper assessment of hazards that may affect the building-related systems.

A.2 Understanding vulnerabilities, risks and threats

To identify cyber-security threats to building data and building systems, the relationships between risks, vulnerabilities and threats need to be understood. Figure A.1, which is adapted from ISO/IEC 15408:2009, illustrates the elements of risk and their relationships. The starting point is that an owner (or user) places some value on an asset, which may be a physical asset or an information asset. This value is derived from some benefits provided by the asset, including its continuing existence and use. Benefits may include compliance with specific legal or regulatory requirements, for example, Basel II for the financial sector with regard to operational risk arising from buildings and building systems, and the Sarbanes-Oxley requirements for risk disclosure that apply to public companies in the United States of America.

Figure A.1 – Relationship of risk elements [adapted from Figure 2 in ISO/IEC 15408:2009]

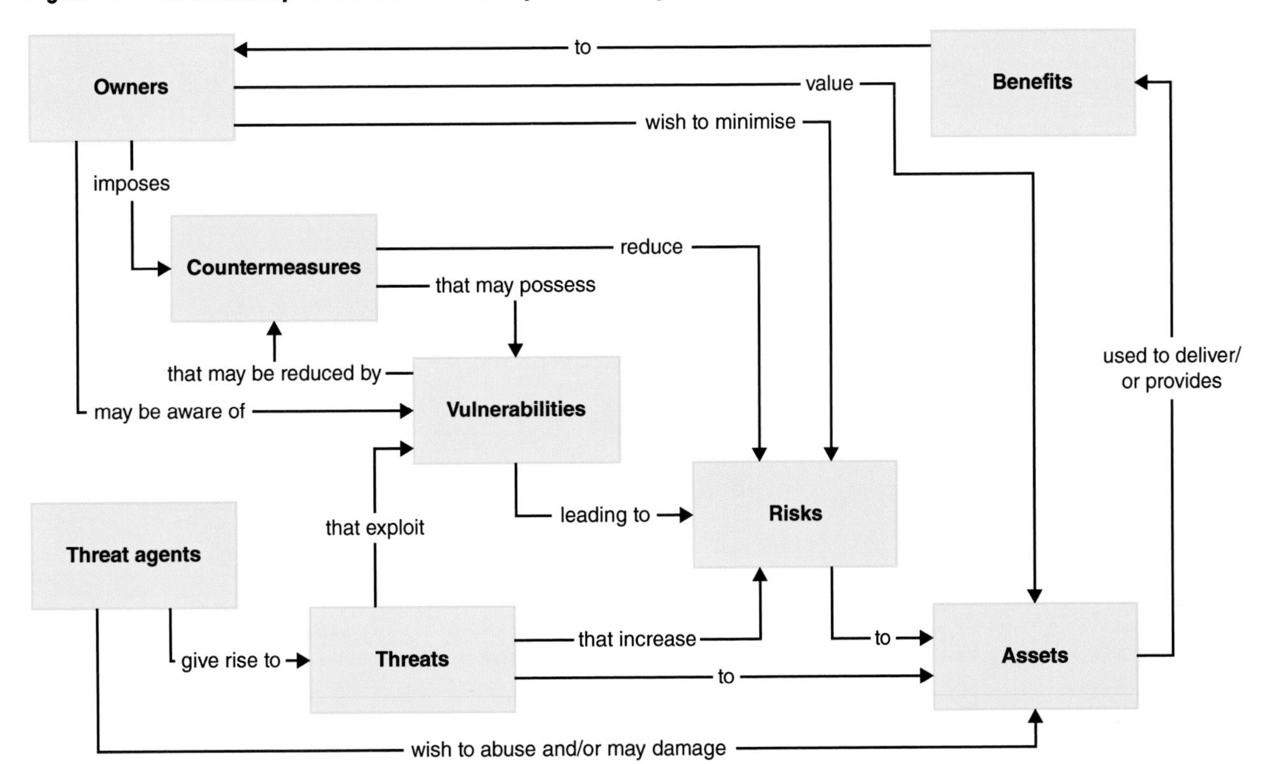

An asset may have a vulnerability, the exploitation of which leads to the risk that the value of the asset may be affected. The arrival of a threat agent that can exploit the vulnerability gives rise to a threat to the value of an asset and increases the probability of the risk being realised. When the owner becomes aware of a threat they may wish to minimise the risk by imposing countermeasures. However, the owner should be aware that the use of countermeasures may themselves introduce additional vulnerabilities. For example, if disk encryption is used to protect data to ensure confidentiality on portable devices and a user then forgets the password or key, the data stored on the device will no longer be available, resulting in loss of possession or control. Obviously this risk can in turn be reduced by the imposition of further countermeasures, for example, automated backup of data to a secure storage area on one of the organisation's servers. The key to managing cyber-security risks is the understanding of the vulnerabilities and potential threat agents, and addressing those risks that have an unacceptable likelihood or impact.

Example: vulnerabilities, risks and threats

An owner values a building because of the rental income derived from the tenants that occupy it. The building has a building management system that is connected to the internet and the system has a default password set on the administrator's account. This is a vulnerability that could be exploited by a hacker resulting in damage or disruption to the building systems. There is a risk that this may lead to loss of rent from the tenants.

As countermeasures, the building owner might address the problem of the default administrator's password by establishing a policy that the password should be changed on a regular basis and by setting minimum standards for password complexity. The account could still be vulnerable if a weak password was chosen or it became compromised. An additional countermeasure, the introduction of two-factor authentication (for example, the use of a security token) could reduce the risks associated with the choice of weak passwords and/or password compromise.

Performing a vulnerability and risk analysis will require an understanding of:

(a) how the building and its systems support its operational use;
(b) the criticality of areas within the building and any building services that support or protect them;
(c) the relationships between building systems;
(d) the relationships between building systems and associated business processes;
(e) the dependencies of buildings systems on infrastructure, for example, utilities;
(f) the technical composition of building systems, in terms of hardware and software components and the builds or revisions that are being used;
(g) the nature of potential threat agents and their action with regards to the building and building systems; and
(h) the presence and permeability of any secure perimeter that prevents or limits access to the building and its associated utilities, plant and machinery.

This is a complex information set that will never be static. Any risk or vulnerability analysis therefore represents a snapshot at a particular instance, where the assessment of risk or vulnerability affecting individual assets may change dramatically with the emergence of a new vulnerability.

Example: software defect creating a new vulnerability

When the Heartbleed software defect was discovered the initial reports were that it only affected the use of SSL on some websites. It was subsequently reported that the affected OpenSSL software was in use by a number of major manufacturers of industrial control systems. Until the manufacturers had developed and the systems maintainers had deployed suitable patches to correct the defect, affected systems were vulnerable to attacks on any SSL sessions. The problem was potentially exacerbated by the ease with which Shodan, a specialist search engine, could be used to identify potentially vulnerable systems.

A.3 Cyber-security attributes

The built environment involves a variety of technologies, existing and emerging, and the cyber-security approach adopted will vary from building to building, depending on the complexity, ownership, use and the supply chain supporting the design, construction, operation and occupation of the building. In the built environment, cyber security is therefore best addressed by considering a set of security attributes, thus allowing appropriate solutions to be adopted, based on the nature of the building and potential threats.

The key attributes[4] of cyber security as applied to cyber–physical systems are outlined below. When considering these attributes, a risk management approach should be adopted, which will inform the degree to which any preventative or protective measures are implemented and the degree to which any residual risk is accepted.

---

[4] The first six attributes are based on the Parkerian Hexad, developed by Don B Parker in 1998.

(a) *Confidentiality*
The building system and associated processes shall be designed, implemented, operated and maintained so as to prevent unauthorised access to, for example, sensitive financial, medical or commercial data. All personal data shall be handled in accordance with the Data Protection Act and additional measures may be required to protect privacy due to the aggregation of data, information or metadata.

(b) *Possession and/or control*
The building system and associated processes shall be designed, implemented, operated and maintained so as to prevent unauthorised control, manipulation or interference. An example would be the loss of an encrypted storage device: there is no loss of confidentiality as the information is inaccessible without the encryption key, but the owner or user is deprived of its contents.

(c) *Integrity*
The building system and associated processes shall be designed, implemented, operated and maintained so as to prevent unauthorised changes being made to assets, processes, system state or the configuration of the system itself. A loss of system integrity could occur through physical changes to a system, such as the unauthorised connection of a Wi-Fi access point to a secure network, or through a fault such as the corruption of a database or file due to storage media errors.

(d) *Authenticity*
It shall be possible to verify the authenticity of inputs to, and outputs from, the building system, its state and any associated processes. It shall be possible to verify the authenticity of components, software and data within the system and any associated processes. Authenticity issues could relate to data such as a forged security certificate or to hardware such as a cloned device.

(e) *Availability*
The building system and associated processes shall be consistently accessible in an appropriate and timely manner. In the event that either a system or associated process suffers disruption or impairment, or an outage occurs it should be possible to recover a normal operating state in a timely manner. A loss of availability could occur through the failure of a system component, such as a disk crash, or from a malicious act such as a denial of service attack that prevents the use of a system connected to the internet.

(f) *Utility*
The building system and associated processes shall be designed, implemented, operated and maintained so that the use of their assets is maintained throughout their lifecycle. An example of loss of utility would be a situation where data has been encrypted and the encryption key has been lost or forgotten; the data still exists, there is no loss of possession or control, but it is unusable.

(g) *Safety*
The design, implementation, operation and maintenance of a building system and its associated processes shall take the above attributes in to account so as to avoid jeopardising the health and safety of individuals, the

environment or any associated assets. A safety issue could arise through malware causing a failure to display or communicate building systems alarm states, for example, the failure of a smoke detector or other sensors that results in damage to property or loss of life.

A.4 Cyber-security context

A building does not exist in isolation: its resilience and cyber-security requirements will be determined by factors such as its location, nature of its use, its occupier and the threat agents that could affect it. To establish the requirements for cyber-security measures, and the subsequent management of those measures, it is recommended that an analysis is conducted to understand the building's context by addressing the factors below and their impact on the attributes listed in the previous section.

(a) *People*

The interaction of people with the building systems needs to be understood, taking into account both how an individual is impacted and influenced by the building, and how an individual impacts and influences the design and evolving nature of the building. Some questions to ask during the assessment include:

i. who needs access to the building data and systems?
ii. what types of access are required to the building data and systems?
iii. how is this access provided, and is it only required within the building?
iv. what access controls will be required (for example, can an individual create, read, update or delete the building data, and what level of control does an individual have)?
v. are access privileges regularly reviewed to ensure that individuals' privileges are consistent with their job roles and functions?
vi. are system access logs regularly reviewed and anomalies investigated?

These questions may establish, for example, that the building management systems need to be accessible by the technical facilities management team from within the building, from an off-site out-of-hours monitoring centre and by the HVAC maintenance personnel (both on site and remotely).

(b) *Awareness and understanding*

The training level and needs of the stakeholders involved in the specification, design, implementation and operation of building-related systems need to be understood. Some questions to ask during the assessment include:

i. what level of cyber-security awareness and understanding is required by individuals?
ii. do individuals understand the organisation's policies, processes and procedures for the creation, use and maintenance of building data?
iii. do individuals understand the organisation's policies, processes and procedures for the operation and maintenance of building systems?
iv. are processes and procedures in place to update individuals about any changes in policies, processes and procedures?
v. are individuals briefed in a timely manner on changes in threats, risks and the required countermeasures?

vi. are contractors, temporary and agency staff provided with cyber-security awareness training as part of their induction?

These questions may establish, for example, that all staff and contractors are given cyber-security awareness training as part of their induction briefing and are regularly briefed on changes to building-related cyber-security policies, processes and procedures.

(c) *Information and data*
The information and data that are created, used and/or processed by the building systems and associated processes needs to be understood. Some questions to ask during the assessment include:

i. what information and data, including sensor data, do the building systems require to function?
ii. how is this encoded?
iii. how and where is it stored?
iv. what would the consequences be if the data were lost and therefore no longer available?
v. who owns the data?
vi. how is it made available and what restrictions are there on its use?

These questions may establish, for example, that the building's cyber-security policy requires the building manager to minimise the amount of personally identifiable information stored and processed within building systems, to ensure that there is a process in place to review data held and to securely remove any unwanted data or information.

(d) *Electromagnetic spectrum*
The channels used by the building systems both to communicate between systems and/or sub-systems, and used by sensors to measure and actuators to control, can all be vulnerable to attacks and interference. Some questions to ask during the assessment include:

i. what channels, technologies and parts of the overall spectrum are used to communicate and share building data between building systems and with any users who need to access or use it?
ii. what channels, technologies and parts of the electromagnetic spectrum are used to control and integrate building systems?
iii. to what extent are the communications confined to the building, and will remote access to, or remote processing of, communications be required?

These questions may establish, for example, that the building management systems primarily use a dedicated wired LAN, but within the facilities management offices there is connectivity to end user devices (laptops, tablets, etc.) using Wi-Fi. Within the plant rooms there is some use of Bluetooth to provide local control of HVAC sub-systems.

(e) *Building systems*
These systems may exist within the building, be located in the immediate area (for example, as part of a campus) or be remote. Some questions to ask during the assessment include:

i. what is the totality of building systems that are involved in the creation, use, maintenance, storage and transmission of building data?
ii. to what extent are these systems dedicated to one specific building?
iii. are the building systems shared by different activities?
iv. are the systems accessible by any third parties, either within or outside the building?
v. what are the typical operating lives of the building's systems?
vi. for any existing building systems, how long before they become unsupportable, obsolete or need to be replaced for business and/ or operational reasons?

These questions may establish, for example, that the dedicated building management system is connected to the individual HVAC control systems located within the building, it has no connectivity to the building occupier's enterprise systems.

(f) *Infrastructure*
The infrastructure supporting the building will include the supply of utilities (energy, water, telecommunications etc.) and physical assets (cabling, pipes, ducts, risers, sumps, etc.) that form part of the building or its surroundings. Some questions to ask during the assessment include:

i. what physical and electronic infrastructure is used to create, access, process and store building data, including any communications and networking components?
ii. to what extent is this infrastructure dedicated to building systems or is it shared with different activities?
iii. to what extent is this infrastructure shared with third parties?
iv. what dependencies does the infrastructure have on other critical services or infrastructure?
v. are there any critical supplies required to ensure the ongoing operation of building systems and any processes or services they support?

These questions may establish, for example, that the building management system is located in a secure plant room and is connected to the HVAC control systems by a dedicated LAN running over its own cables in spaces and cableways managed by the facilities management team.

(g) *Environment*
This factor considers the wider social, political and legal aspects that are pertinent to the building and building systems, for example, the presence or absence of specific legal or regulatory requirements that concern the design and operation on a building. Some questions to ask during the assessment include:

i. what societal, technological, economic, environmental, political, legal, ethical and demographic (STEEPLED) considerations are associated with the creation, use, management and exploitation of building data?
ii. what are the STEEPLED considerations associated with the design, implementation, operation and maintenance of building systems?

iii. what are the STEEPLED considerations associated with the use of the building, i.e. its ownership and function and impact on users and/or visitors?

iv. is the building data processed, stored and used within a single jurisdiction or are multi-national jurisdictions involved?

v. are there specific financial or regulatory requirements underpinning the design and operation of the building systems?

vi. are there specific regulatory or legal reporting requirements that could result in risks to building systems?

These questions may establish, for example, that the building and its data are located within a single jurisdiction. The access control system that contains personal identifiable information is a separate system and is not connected to the building management system, the HVAC control systems or the network and communications infrastructure they use.

A.5 Potential threat agents

A.5.1 Types of threat agent

Threat agents may be classified into one of three categories:

(a) Natural – these threat agents are related to solar, weather, animals or insects. Their impact on building systems may result in a failure or significant impairment of one or more of the utility supplies or the building systems. Depending on the threat, building systems may be disrupted or damaged, and building data collection may be lost or corrupted.

(b) Friendly – these threat agents are not seeking to harm the building systems or data, but may access the systems without the permission or knowledge of the building systems' owner or may cause accidental damage. This group include: researchers, ethical hackers, security authorities, law enforcement and the military. Their motivation is generally to investigate weaknesses and vulnerabilities in systems, for example, security researchers undertaking scans of systems connected to the internet, to compile statistics on open network ports and systems software in use.

(c) Hostile – these threat agents have varying capabilities and motivations. They emanate from two groups, which are not mutually exclusive, i.e. a particular cyber-security threat could be the work of a hostile outsider, knowingly or unknowingly supported by an insider. The groups are defined as follows:

    i. outsiders – this is a person or persons unconnected with the building owner, the building occupier or supporting contractors – in essence, a person who does not, by right, have privileged access to the building, the building systems or the building data.

    ii. insiders – this is a person or persons connected with the building owner, the building occupier or supporting contractors – in essence, a person who has been granted some level of authorised or privileged access to the building, the building systems or the building data and puts their privileged access to a use that is not intended or allowed. A threat agent with enhanced IT skills may, through social engineering, influence a disaffected employee by, for example, giving detailed instructions or by providing software or equipment to install on the building's network.

The hostile threat agents are examined in more detail in the next section of this appendix.

When considering the potential threats to a building the assessment should recognise that the threat is not only to the physical building and the systems it contains, but also to those elements of the building systems or building data located outside of the building. These could include the use of cloud storage of data, use of cloud or software as a service to manage building processes, and the BIM environment, models, data, etc.

A.5.2 Potential hostile threat agent groups

A building could come under threat from a wide variety of hostile threat agents, either directly targeted, or indirectly due to malicious activity targeted elsewhere. The purpose of the list below is to help identify potential types of malicious threat.

(a) An individual – the severity and sophistication of the threat will be determined by the individual's capabilities, for example:
   i. Negligent or careless employee – fails to follow acceptable use or other security policies and compromises system security;
   ii. Disaffected employee with limited IT skills – motivations will vary; the intent may be to steal or leak sensitive information, to sabotage or disrupt building occupancy or operations, etc. The amount of damage they can inflict will depend on their role, system access rights and the efficacy of cyber-security measures related to the building systems and data.
   iii. Disaffected employee with significant IT skills, including system administrators – these individuals can do significant damage, particularly if they have wide ranging systems access with administrative privileges. They may have sufficient knowledge and ability to bypass controls and protective measures, and may be adept at removing evidence of their activities, for example, deleting or modifying entries in system logs.
   iv. Script kiddies – individual hackers with limited knowledge who use techniques and tools devised and developed by other people. The ready availability of hacking and denial of service tools on the internet (and in some cases distributed with technical magazines) means that the level of technical understanding required to launch an attack has been significantly reduced.
   v. Cyber vandals – this group can be very knowledgeable and may develop or further expand their own tools. Their motives are neither financial nor ideological – they carry out hacks or develop malware because they can and want to show what they can do. They may, for example, deface a website or break into a server to steal user credentials, which are then posted on a public website to demonstrate their ability.

(b) Activist groups – often referred to as hacktivists, these groups comprise ideologically motivated individuals that may form dynamic groups or sub-groups. Their actions are effectively online protests, which may have the aim of disrupting systems, or acquiring confidential or sensitive information for publication or dissemination so as to embarrass their target(s). The impact of small activist groups can be significantly magnified when, as some groups have demonstrated, they recruit or persuade naïve third parties to join in by allowing the installation of malicious software on the recruits' computers, thus creating botnets and magnifying the effect of any distributed denial of service (DDoS) attacks.

(c) Competitors – this group is typically large corporations seeking to create competitive advantage. They may act directly or through third parties, with the aim of harming a rival by collecting business intelligence, stealing intellectual property, gathering competitive intelligence on bids or disrupting operations to cause financial or reputational loss. Depending on size, sector, geographic location and the sophistication of a large corporation's cyber capabilities they may be able to perform sophisticated malicious activities to target and infiltrate their victims.

(d) Cyber criminals – sophisticated criminal groups, perpetrating a wide range of illegal IT-enabled crime. The motivation is to profit from illegal activities, and their focus has mainly been on fraud, thefts from accounts and theft of intellectual property. However, cyber-criminal activities also include blackmail and extortion through use of malware to encrypt data or threats of denial of service attacks on corporate websites. In respect of buildings, cyber criminals may seek to intercept or access BIM models related to sensitive locations, for example, layouts and security information for prisons, pharmacies and buildings that handle or store cash or other valuable items as a precursor to a physical attack on these premises. The sophistication of the malware used by these groups is increasing and there is evidence of a cyber-crime market, where developers, providers and operators create, supply and operate sophisticated malware and cyber-crime tools on a commercial basis, making their tools available to third parties.

(e) Terrorists – they are becoming increasingly IT aware, and already make extensive use of the internet to distribute propaganda and for communications purposes. Well-funded groups could take advantage of the service offered by cyber criminals, seek support from a nation state or encourage internal members to adopt these methods of attack. With the widespread use of electronic- and computer-based technologies in the built environment, terrorist groups could rely on the various toolkits available for download to disrupt or damage buildings by attacking building systems. Terrorists may also exploit poorly secured building data to enable remote hostile reconnaissance of targets, thus reducing the time they need to spend in or near a target building or site.

(f) Proxy terror threat agent with nation state support – this is effectively state-sponsored terrorism, where the proxy party is used to provide deniability. This type of group effectively has the capacity and sophisticated technical support available to a nation state made available by the sponsoring nation. This group could include cyber fighters, i.e. groups of nationally motivated individuals who threaten or attack other groups, businesses and the infrastructure of other nation states. The cyber fighters may be seen as a type of hacktivists, but their interest is the support of a nation state and as such they may enjoy significant sophisticated technical support from that nation state.

(g) Nation states – it is acknowledged that some nation states are actively involved in cyber-attacks on a wide range of organisations to acquire state secrets or sensitive commercial information and intellectual property. They may also threaten the availability of critical infrastructure in other nation states. During periods of heightened international tension and conflict, these activities may include more widespread attacks as evidenced by malware such as Stuxnet, Duqu and Flame.

When considering the potential threats from the hostile groups listed above, it is important to recognise that there may be some convergence between the aims and objectives of individual groups. For example, some of the malware developed by cyber-criminal gangs includes sophisticated command and control functionality, allowing secure exfiltration of information and update of modular components to deliver new or varied exploits over time. Thus a machine or device that was compromised initially for financial crime could be used in future to access sensitive data or to provide a backdoor to allow attacks on building systems.

It would be wrong to assume that simply because the building is not occupied by a government body or high profile commercial target that those building systems and associated business processed will not be affected. The cyber-attack on the Target stores group in the United States started with an attack on their HVAC maintenance supplier and resulted in the loss of credit card details for over 140 million people. Denial of service attacks have, on occasion, severely disrupted the internet, resulting in very slow traffic handling; this would be a serious issue if real-time elements of building control were outsourced to a cloud service provider.

The 'computer glitch' in August 2013, which resulted in the unplanned release of cell door locks in a Miami state prison[5], is an extreme example of how a cyber-security incident could have significant business impact. Whilst this was said to be a computer malfunction, a poorly designed system architecture or poor implementation could easily expose systems like this to attacks from malicious threat agents.

A.6 Enterprise and industrial control systems architecture

Achieving cyber security in the design and operation of systems is not just about the hardware and software elements, there are also considerations about the location of components, the business processes, and data and information requirements. The overall enterprise and control systems architecture therefore affects the risk profile of a building. A potential cause of serious cyber-security breaches is the failure of an organisation to appreciate the structure and operation of its systems and associated business processes. This can result in a number of undesirable situations, including:

(a) accidental or inadvertent exposure of sensitive systems, applications or data to unauthorised users;
(b) loss of resilience or redundancy where the criticality of components or processes are not understood; and
(c) emergent failure modes that result in the cascade or catastrophic failure of critical systems or processes.

In following the processes described in Appendices B, C and D to develop the cyber-security strategy, policies, processes and procedures, and to assess and manage the cyber-security risks, it is important that the enterprise and control systems architecture is understood. In particular an understanding of any relationships between the enterprise systems and the control systems is critical. If a systems architectural model is not available, one should be developed. Guidance on the development of enterprise and systems architectures can be found in the EACOE Enterprise Framework and BS ISO/IEC 42010 respectively.

A.7 Standards, guidance and good practice

There is a wide range of security-related standards, guidance and best practice available that apply to IT and industrial control systems. Appendix I lists a broad

[5] For further information, see http://www.wired.com/2013/08/computer-prison-door-mishap/.

range of such documents. Much of the material is written from an information systems security perspective and needs to be carefully interpreted when applying it to systems in the built environment. For example, the application of some security techniques to safety critical systems may hinder their operation in an emergency situation. A complexity that will increasingly occur in the built environment is that of the integration of safety critical alarm and control systems with conventional enterprise and office IT systems. These situations need to be carefully managed by the building owner as the office elements may operate under security policies and procedures originating from ISO 27000 series documents, whereas control and safety systems are more likely to operate under regimes determined by the IEC 61508 and IEC 62443 standards.

APPENDIX B

Developing a cyber-security strategy

Developing a cyber-security strategy for a building so that it achieves the desired level of cyber hardening may require a collection of:

(a) tools;
(b) policies;
(c) security concepts;
(d) security safeguards;
(e) guidelines;
(f) risk management approaches;
(g) actions;
(h) training;
(i) best practices; and
(j) assurance and security technologies.

The nature and composition of this collection will need to vary over a building's lifecycle, responding to changes in the threat landscape, the building's use and profile, and as a consequence of technology developments. Further details on developing a cyber-security strategy for a building are provided in Appendix C. The following sections outline the key aspects of the approach that should be taken by the building owner. Where the building is not owner occupied, then the occupiers (for example, lessees) should read the document.

B.1 Overall development process

To develop a cyber-security strategy for a building and/or the BIM environment, the building owner needs to:

(a) understand the context;
(b) analyse and evaluate the risks; and
(c) manage and mitigate those risks.

They cannot be treated in isolation, as a change in context may affect the risk assessment, which in turn may require additional measures to be taken to mitigate or eliminate the risk. As the building's context will change over time, the cyber-security strategy will need to be periodically reviewed. The frequency of reviews should be determined by the nature of the building's use, the nature of the building occupier and by the general business and political environments.

The cyber-security strategy should have a central role in informing the development of the enterprise architecture (which encompasses business goals, processes, materials, roles, locations and events)[6] and the design, implementation and operation of the cyber-security policy, processes and procedures. The strategy needs to focus on the high level building and business impacts, i.e. what is of

---

6 Based on the EACOE Enterprise Framework – http://www.eacoe.org.

strategic value and needs protecting. Issues that cause minor disruption and can be handled in a tactical manner should not be the strategic cyber-security drivers for the building systems.

The process for developing or updating the cyber-security strategy is illustrated in Figure B.1. It starts with the activities required to understand the context. This involves understanding what assets and benefits are at risk and may need protection. It also involves the identification of the risks and the nature of the associated threats and vulnerabilities. The presence or absence of a threat agent that may exploit a vulnerability will have a bearing on whether a risk needs to be mitigated. However, individual vulnerabilities should not be ignored simply because there is no known threat agent that may exploit them. Such an agent may emerge in the future and be able to take advantage of serious unmitigated vulnerabilities. The process shown in Figure B.1 should form part of an ongoing monitoring and review process that takes into account business developments and learnings from the operation of building systems and associated business processes.

Figure B.1 – Information required in a cyber-security strategy development process

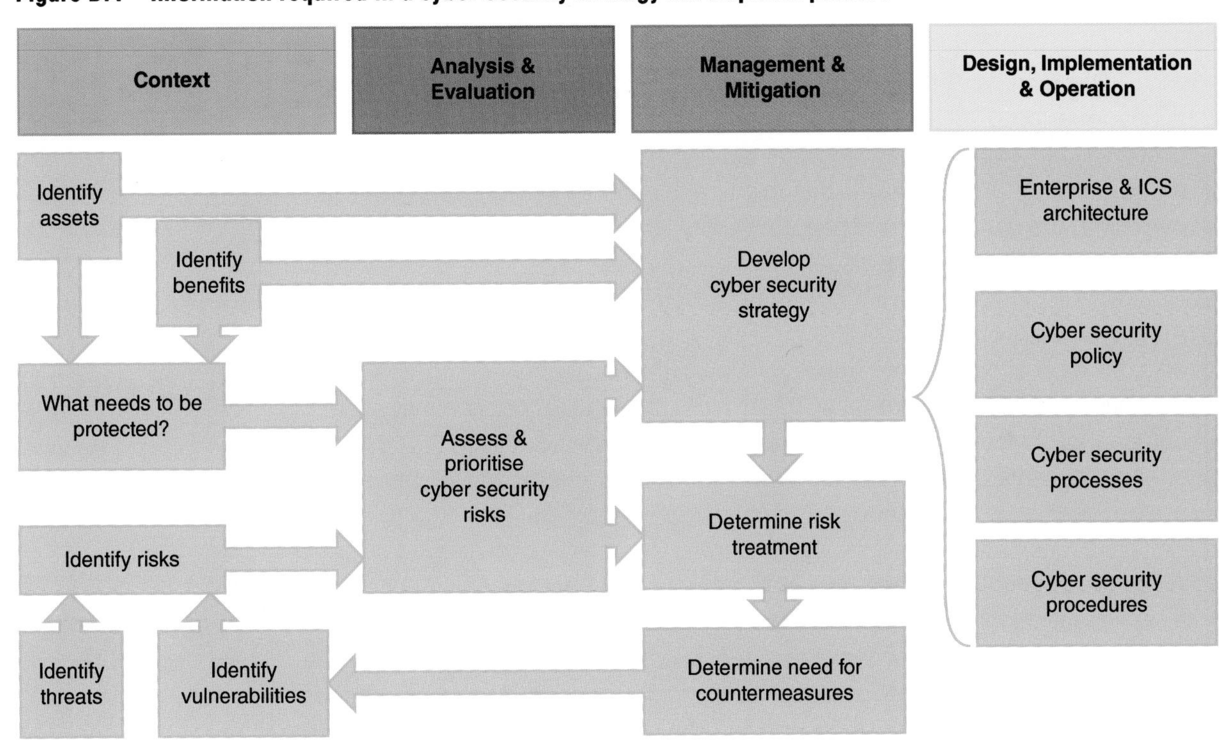

Once the risks and the assets and benefits requiring protection have been identified, an analysis and evaluation exercise should be undertaken. This analysis should consider the likelihood and impact of individual risks and combinations of risks occurring so as to determine the potential exposure or consequences in terms of operational, financial and reputational damage, loss and impairment. By understanding the exposure or consequences, the risks can be ranked and costs assessed for managing individual and/or groups of associated risks using appropriate countermeasures.

The next stage is to determine the risk treatment, taking into account the exposure and the costs associated with any countermeasures. When determining the risk treatment, four options are typically considered:

(a) avoidance – this may be by deciding not to start or continue with the activity that gives rise to the risk, for example, choosing not to build or to dispose of a particular asset;

(b) reduction – by taking steps to reduce either the impact of the risk or the likelihood of it occurring, for example, through controls and protective measures;

(c) sharing the risk with another party or parties – this may be achieved through contracts, outsourcing or insurance; and

(d) retention – accepting the risk and taking appropriate steps to manage the consequences, for example, budgetary provision or business continuity measures.

There may be some interaction between determining the risk treatment and determining the countermeasures. For example, the intention may be to transfer the risk to a third party, but during negotiations the third party only partially accepts the risk. This may require the building owner or occupier to implement additional countermeasures to achieve the required degree of mitigation.

Example: risks affecting support to an access control system

A building owner has decided to invest in a new access control system for a headquarters building. The decision is based on a business case that assumes that savings may be obtained by reducing the manned guarding of the building. It is proposed that this is achieved by relying upon smart ID cards and the use of PIN numbers to identify visitors holding building passes, thus removing the need for visual checks on passes.

As the smooth operation of the access control system is essential for the efficient operation of the building, the building owner determines that continuous operation is required, with high availability, i.e. minimal system outages for maintenance or repair. The system supplier advises that this can only be achieved if the support personnel have remote system access to perform system maintenance and to investigate problems.

A risk assessment of this remote access functionality determines that it poses a significant risk of being hacked. If the system was hacked, the attacker could disable the system or grant unauthorised access to the building. To mitigate this risk, the building owner decides, as a countermeasure, to implement two-factor authentication (for example, the use of a password and a security token) and will provide suitable security tokens to all support personnel who need to access the system. This reduces the risk from a malicious outsider hacking into the systems. However, there remains a residual risk that one of the support personnel may pose an insider threat. The building owner and supplier agree to manage this risk through improved personnel security measures.

This overall process will need to be repeated when there are significant changes to the building, its use, and the building systems including their interconnectivity and exposure to attack. There is a need to maintain situational awareness, i.e. to understand developments in the nature of cyber-security threats and vulnerabilities. Given the rapidly changing nature of cyber-security threats, and depending on the building and its use, it may be appropriate to review the strategy on a periodic basis, to ensure that new threats and vulnerabilities have been appropriately addressed. The frequency of the reviews may be determined by the changes identified as part of any situational awareness activities undertaken by the building systems owner.

Example: good resilience reduces risk

Whilst determining the risk treatment, it may be appropriate to consider a Plan B, i.e. a strategy to deal with the occurrence of a cyber-security incident and a failure in countermeasures. When examining or developing a Plan B it is important to make realistic assessments of how long it is likely to take to repair or restore affected systems or services and any consequential impact on the assets and benefits. These recovery aspects should be part of a business continuity plan for the building owner/occupier. The combination of any deployed countermeasures and a Plan B will determine the risk exposure.

B.2 Understanding the context

The starting point for developing or updating a cyber-security strategy for the building is to understand the context. As illustrated in Figure B.1, this involves identifying:

(a) assets, which may be:
   i. tangible, for example, the building or items stored within it; or
   ii. intangible, for example, a BIM model, both in terms of the intellectual property invested in the design as the model develops, and its use in long term asset management of the building.

(b) benefits, which may be:
   i. financial, for example, growth in capital value, rental income, savings in capital and operational expenditure derived from the information asset (such as the BIM model);
   ii. regulatory compliance, for example, Basel II (in respect of the operational risks to banks arising from damage to physical assets, business disruption or systems failures);
   iii. operational, for example, providing space or accommodation for specific activities; and
   iv. societal, for example, enabling delivery of healthcare or education.

(c) threats, which may be a person, thing or event that could cause damage, danger or disruption, as described in Appendix A.5.

(d) vulnerabilities, which will arise from specific susceptibilities and may result in the loss of a combination of authenticity, availability, control, integrity, safety, security or utility.

This identification of assets and benefits can be achieved by examining the building and its use. It is important to consider the use from a range of stakeholder perspectives, for example, a hospital delivers healthcare to its patients and employment benefits to its staff.

The identification of threats will need to take into account the location, ownership, occupation and use of the building. Threats from nature should not be ignored, for example, the impact of severe weather causing flooding or physical damage either to the building or in a way that prevents staff, users or critical supplies from getting to the building. It should not be assumed that because a particular natural event has a probability of occurring once in a hundred years that it will never happen. There has been a number of such 'one in a hundred year events' occurring over the last decade.

To identify vulnerabilities, a comprehensive vulnerability analysis should be undertaken by the building owner. This analysis addresses the cyber-security attributes, described in Appendix A.3, and the requirements analysis, described in Appendix A.4. The interaction between the attributes and the analysis elements, as illustrated in Figure B.2, can be quite complex. The intersections between horizontal and vertical axes in this diagram represent specific areas to be considered. Appendix J provides examples of questions to consider when exploring the interaction between individual attributes and the analysis elements. This process needs to be completed in sufficient detail to understand the threats and vulnerabilities that need to be addressed at a strategic and policy level. The context can be understood once this analysis has determined what needs protecting and the nature of the risks.

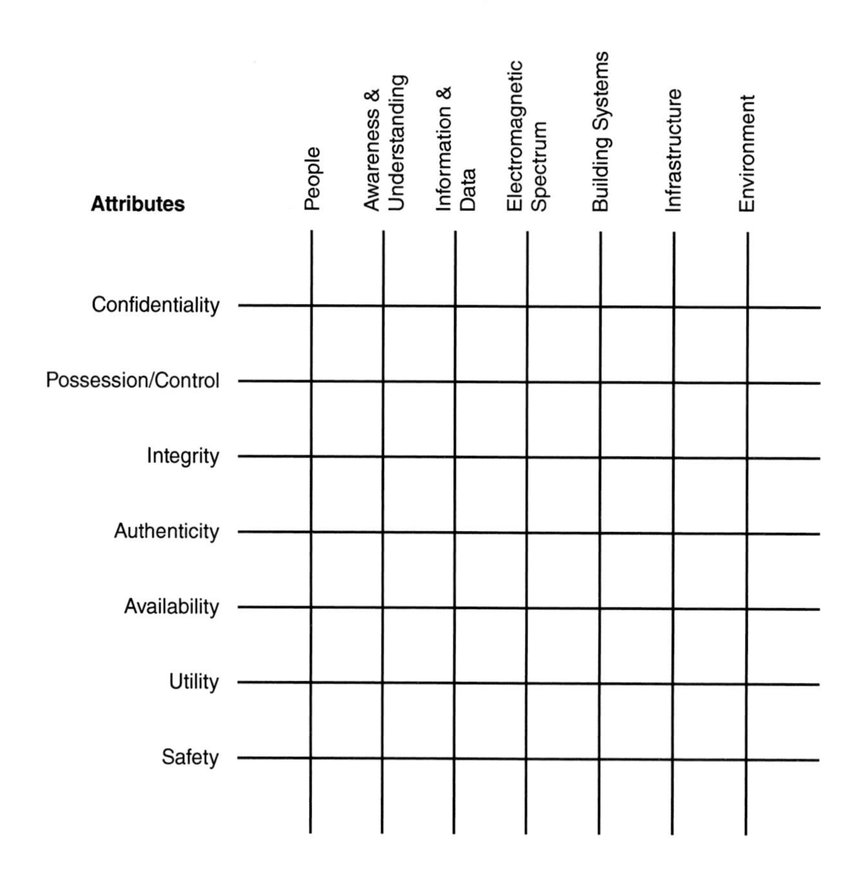

Figure B.2 – Cyber-security attributes interaction with analysis elements

Example: interaction between confidentiality and the human element

This interaction relates to the protection of confidential building data across the building lifecycle. The weakest link in any security system is often the human who, accidentally or maliciously, causes disclosure of confidential information. To understand the potential vulnerabilities the analyst needs to know what building data is confidential; this may include intellectual property related to innovative design or operational features, commercially sensitive data, personally identifiable information, etc. The analysis will need to address questions such as how and where the data is accessed by people, where it is processed and stored, the access control requirements, etc.

B.3 Cyber-security risk analysis and evaluation

The first stage in this process is to understand the context and undertake a risk analysis and evaluation. This is described in more detail in Appendix D.1. The methodologies used, for example, quantitative and/or qualitative analysis, and the level of detail involved will depend on the criticality of the building and its use.

A weakness of both quantitative and qualitative risk analysis methods is that they are likely to ignore or understate new or novel cyber-security risks, particularly those relating to advances in building-related information and communications technologies. To overcome these weaknesses the risk analyst should consider using some of the techniques that are often applied to safety critical systems, for example, failure mode and effects analysis (FMEA) or anticipatory failure determination (AFD) prediction. The latter technique employs a method by which potential failures are identified not by asking what might go wrong, but by asking how someone can make it go wrong and how that failure could be prevented or mitigated. This encourages the generation of scenarios from combinations of single failures that might have greater impact than individual failures. Using these engineering techniques can help to both identify potential vulnerabilities and to explore the consequences of a number of risks that may interact at one time.

B.4 Managing and mitigating cyber-security risks

Cyber-security management should be based on a clear strategy and typically comprises a range of activities and controls that an organisation implements to ensure that it is effectively managing its cyber-security risks. The activities and controls cover four aspects: policy, processes and procedures, people, and technical. A specific risk may require a number of activities and controls to reduce the likelihood and consequences of a risk to an acceptable level. However, through situational awareness activities, it may be necessary to adapt these to address emergent threats.

Example: handling a specific risk

An organisation has determined that to manage the risk of unauthorised access to an online account (i.e. an account that is accessible from the internet), a series of measures should be put into place:

(a) policy – implementation of a policy that these accounts must use two-factor authentication (2FA) and limiting the type of accounts accessible online to those that do not have system administrator level privileges.
(b) processes and procedures – specification of the strength of passwords, the mechanism for, and frequency of, password changes, and the prohibition of reusing or sharing passwords.
(c) technical – the measures required to enforce the password strength rules and frequency of change rules, and the implementation of 2FA components;
(d) people – provision of user education and awareness training for all personnel with accounts that are accessible online to ensure that the users understand the risks, how to comply with the policy and the operation of the processes and procedures.

The above measures have been put into place, when as part of a routine situational awareness review, the building systems manager becomes aware that there is a major security flaw in the 2FA system. These weaknesses are apparently already being exploited by hackers. Following a risk review it is decided that this represents an unacceptable level of risk to the organisation. Use of online access to these accounts is suspended pending the completion of a project to upgrade the 2FA system to address the flaw.

This example reflects a strategic approach that had been adopted to enable online access to some accounts with a tactical response to address a specific newly emerged vulnerability in a dynamic threat environment.

APPENDIX C

Developing a cyber-security policy for a building

C.1 Relationship of cyber-security policies, processes and procedures

The successful ownership, occupation and operation of a building requires a systemic approach; effectively, a set of rules relating to its use. The cyber-security strategy described in Appendix A needs to be translated into practical, actionable steps. This can be achieved through the creation of suitable policies, processes and procedures as illustrated in Figure C.1. Whilst these terms are too often interchanged, 'policies', 'processes', and 'procedures' should be three distinct types of documentation. All three will address related subject matter, but at different levels and with different types of content. Each level has a unique purpose that drives the content contained in each type of document. This cascade from the strategy through the policy to process and individual procedures is important as it provides an auditable trail that links specific actions and activities to the overall vision of how the cyber-security risks will be managed and mitigated.

The ultimate responsibility for the strategy and policy lies with the building owner, as it is the owner's asset(s) that may be attacked, however, the occupiers' views and needs must be understood and taken into account when developing or updating the strategy and policy. For example, a high profile tenant or one who will attract the attention of specific threat agents may require the deployment of specific countermeasures. The apportionment of cyber-security risks between owner and occupiers is a matter that needs to be addressed during any contract or lease negotiations to ensure that obligations and responsibilities are clearly understood and documented.

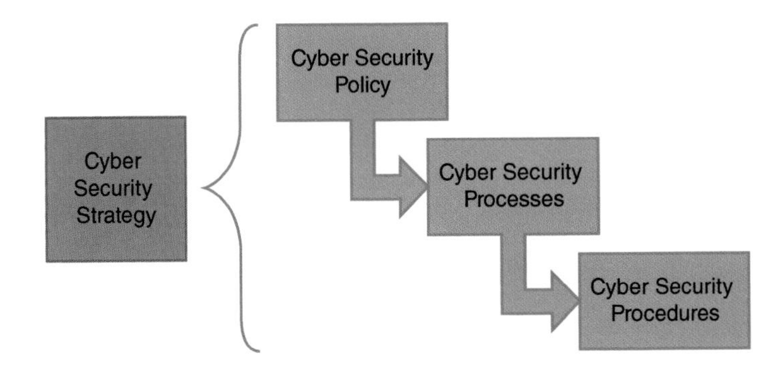

Figure C.1 – Policy, processes and procedures based on cyber-security strategy

C.1.1 Cyber-security policy

The policy will provide the guidelines under which procedures are developed. There is not normally a one-to-one relationship between a policy and a procedure. A policy is not part of a procedure, and a procedure must reflect the business rules contained in the relevant policies. A policy will state what the organisation's policy is, its classification, the line of accountability and who is responsible for the execution and enforcement of the policy, and why the policy is required.

An organisation's cyber-security policy should set out the business rules and guidelines that ensure consistency and compliance with the organisation's strategic direction, including its risk appetite. The policy will lay out the cyber-security business rules under which the organisation and its component parts will operate. Given the cross functional nature of a building's cyber-security strategy, ownership should be at Board level within the building owner's organisation.

The cyber-security policy will relate not only to the building, but to the building systems, associated business processes and building data. For single occupancy buildings it may be aligned to, or part of, the organisation's cyber-security policy. For multi-occupancy buildings it will need to be a standalone policy managed by the building's owner. In both scenarios it needs to be in a form that allows it to be shared with third parties, for example, those responsible for, or contracted to provide, design, construction, operations and maintenance services. This is examined in more detail in Appendix C.2.

C.1.2 Cyber-security processes

A process is a high level view by which the related tasks and their order are identified, so as to deliver a specific service or product. In many cases processes may cross organisational boundaries, for example, departments or functional areas within the organisation, and suppliers outside the organisation. A process should:

(a) indicate where there is a separation of responsibilities and control points and who is responsible for performing the process (for example, organisation, department, division, or team);
(b) what major functions are to be performed; and
(c) how and when the function is triggered.

A process document should cover the whole process, referencing policies and procedures where appropriate, and may include a graphical process map to explain the step-by-step flow. In the context of a building, a cyber-security process may, for example, describe the steps required to set up a new building user on the building's access control system.

C.1.3 Cyber-security procedures

Procedures are the detailed steps required to perform an activity or task as part of a process. A procedure defines the specific instructions necessary to perform the activity or task and may take the form of a work instruction, a desktop procedure, a quick reference guide or a more detailed procedure. Procedures are usually structured by subject (for example, system instructions, report instructions or process tasks). A procedure will usually address only a single task, defining who performs it, the steps to be performed and how the procedure is performed. This decomposition of the process into separate documented components enables them to be compiled into special procedure manuals, for example, those that apply to specific audiences, end users, and for specific purposes.

C.2 Cyber-security policy objectives and scope

The objective of a cyber-security policy for the building, the building systems and building data is to provide management direction and support for cyber-security based on business requirements and any relevant legal, licence or regulatory requirements. The cyber-security policy may be a standalone document or part of an overall building security policy.

The cyber-security policy should be defined and approved by the building owner. If the building is not owner occupied, then the building occupier may be consulted and involved in the development of the policy. Once approved by the building owner, the policy should be communicated to all relevant parties. The scope of the policy should be relevant to the building, its owner and its use. It should set the objectives for managing building-related cyber security . The scope of the policy should cover the design, specification, implementation, operation, maintenance, modification and decommissioning of the building systems, their associated processes and building data.

The cyber-security policy for the building and, where it exists, the building's BIM environment should cover requirements arising from:

(a) the building owner's strategy and, where it is not owner occupied, the building occupier's strategy;
(b) legislation and regulations;
(c) where applicable, licence conditions – this is likely to be required in regulated industries; and
(d) current and projected cyber-security adversities, i.e. the combination of threat and hazard scenarios.

The building's cyber-security policy should contain a statement of the cyber-security objectives and principles that apply to the building systems, associated processes and the building data. These objectives and principles should guide all activities related to the building's cyber security. It should also clearly address the scope of the policy, taking into account both the physical aspects of the building and its location, and the nature and delivery location of IT, communications and business processes that support the asset (building) and deliver its benefits.

The policy should assign responsibility to specific roles; this is particularly important where systems, processes and data cross organisational boundaries. The policy should define the processes for regularly reviewing and updating the policy, and those for handling deviations and exceptions. The policy may be supported by a number of lower level policies that address specific policy topics, for example, the handling of incidents may be addressed in an incident management policy.

C.3 Legislation and building systems

Building systems and building data can be affected by a variety of legislation and regulation. The impact and controls that may be required will vary from system to system. The cyber-security policy may need to address a combination of the following types of legislation:

(a) health and safety – where the failure, misuse or modification of a building system could affect a person's health or lead to injury or loss of life, the system has safety-critical features and appropriate measures need to be taken to protect these features from adverse conditions as a result of a cyber-security incident.
(b) data protection – where there is personally identifiable information (PII) stored or processed in building systems. Examples of where PII may be

used in building systems include the use of biometric data identification and access control, information about any special facilities or support required by individuals with health problems or disabilities; personal data associated with access control systems, etc. The collection, storage and analysis of PII can also have privacy implications.

(c) criminal legislation – depending on the legal jurisdictions in which the building, the building systems and the building data are located there is often legislation relating to unauthorised access to, or use or modification of, computer systems and their data. For example, in the UK the Computer Misuse Act makes it a criminal offence to gain unauthorised access to computer material, to gain unauthorised access with intent to commit or facilitate the commission of further offences, or to make unauthorised modification of computer material. It is thus an offence to use another person's username and password without proper authority to access building data or building systems, to alter, delete, copy or move a program or data, to output a program or data to a screen or printer, or to impersonate that other person using e-mail, online chat, web or other services.

(d) corporate governance – for example, Sarbanes-Oxley. This US legislation came into force in 2002 and introduced major changes to the regulation of financial practice and corporate governance of US organisations, both within the USA and their global operations. Security and governance of risk, particularly those risks with significant financial consequences, are key parts of this legislation

(e) civil legislation – a failure to take appropriate steps to manage the cyber security of building systems, their associated processes and building data could expose any of the parties involved to claims for compensation as a result of negligence, breach of contract, etc.

(f) specific legislation or regulations – some buildings, due to the nature of their use, ownership or occupants, may be subject to additional legal or regulatory requirements, for example, sensitive government buildings may be required to comply with specific national security related regulations. In the UK, sports grounds and stadiums have specific legislation covering their safe use, for example, the Safety of Sports Grounds Act 1975, the Fire Safety and Safety of Places of Sport Act 1987, and the Safety of Places of Sport Regulations 1988. Many of these documents will not contain specific cyber-security requirements, but there is still a need to consider these requirements. A more recent document, Counter Terrorism Protective Security Advice for Stadia and Arenas, does contain advice on information security, but this is primarily aimed at the business and administrative systems rather than the building systems.

Example: specific legislation with cyber-security implications

In *Guide to Safety at Sports Grounds* guidance is provided on the principal means of communication required in a sports ground. It lists eight principal means:

(a) radio communications;
(b) telephone communications (internal and external);
(c) public address systems;
(d) closed circuit television systems (CCTV);
(e) scoreboards, information boards and video boards;
(f) signs;
(g) written communications (via tickets, signs and printed material); and
(h) inter-personal communications.

Of the principal means listed, five are susceptible to cyber-security incidents. Given the extensive use of information and communications technologies in stadia and sports grounds, the operator should be considering the cyber-security risks associated with this critical site infrastructure.

C.4 Cyber-security leadership

Cyber-security leadership requires all levels of management, and not just technical specialists, to demonstrate commitment and support so that cyber-security good practice is integrated into all aspects of a building's systems, associated processes and building data. Strong leadership on the importance of cyber-security matters is likely to engender a positive approach, good awareness and reduce the risk to the organisation and the building.

Key requirements for achieving leadership include:

(a) ensuring that requirements for cyber-security management are included in job descriptions;
(b) recognising and understanding cyber-security issues that may affect the building;
(c) having relevant cyber-security targets;
(d) ensuring that there is appropriate engagement with the workforce, including both employees and contract staff;
(e) allocating appropriate resources and time to addressing building-related cyber-security matters;
(f) ensuring appropriate actions are taken to remedy identified cyber-security issues and that the actions are implemented; and
(g) ensuring good two-way communication with the workforce so that cyber-security issues can be highlighted and discussed and commitment can be obtained to resolve the issues.

An important leadership issue is how to address the management of cyber security across organisational boundaries. These can include the boundaries between IT and facilities management departments, between the owner's or occupier's workforce and suppliers or contractors, and between different occupiers in a multi-occupancy building.

Managing 'process and procedure' aspects

D.1 Cyber-security risk management

In Appendix B, the concepts of cyber-security context, threats and vulnerabilities were examined. The identification and management of cyber-security risks are complicated by the nature of the many systems found in buildings, coupled with constantly evolving threats and vulnerabilities. Figure D.1 illustrates the interaction between elements that affect the residual risks for any building. The building or building system will have a number of vulnerabilities, which may be considered in two groups:

(a) system and systemic vulnerabilities – these lend themselves to be managed using traditional risk analysis methods. System risks typically affect individual systems or the interfaces between them, for example, the failure of a hard drive causing loss of data. Systemic risks affect the building as a whole and may result in cascading failure, for example, loss of power causing all building systems to fail.

(b) dynamic and emergent vulnerabilities – these require the use of situational awareness to identify whether a potential vulnerability may impact the building systems. Dynamic risks may be short term, for example, denial of service attacks by hacktivists on an internet connection. Emergent vulnerabilities will typically arise through the discovery of exploitable hardware and/or software defects, for example, the Heartbleed bug affecting OpenSSL.

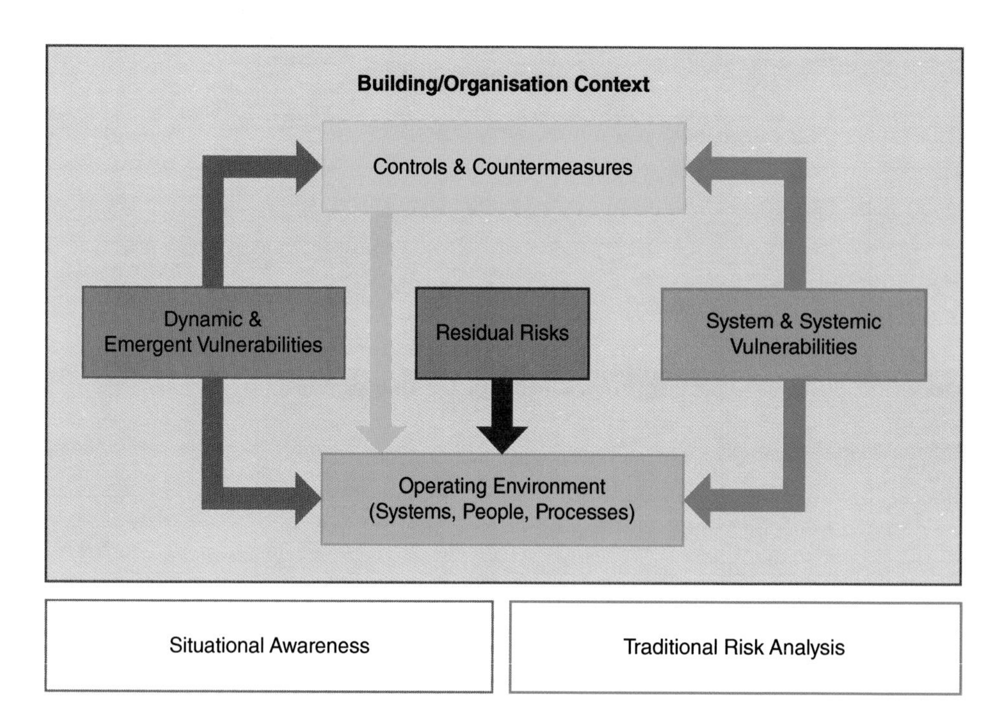

Figure D.1 – Cyber-security risk interactions

The system and systemic risks may be identified and analysed when new systems, processes and organisational structures are being planned and/or implemented, or when changes are being made to existing systems, processes or organisational structures. In addition to any project-related risk review, the risk analyses for building systems and their associated processes should be periodically reviewed and updated over the building's lifecycle. Identification and analysis of these risks is examined further in Appendix D.1.1.

Dynamic and emergent risks are often unknown, hence are difficult to identify using traditional risk management processes. These risks may affect systems (i.e. technology), people or process elements of the building's operations. The complexity of identifying them lies in the diverse and unpredictable nature of these vulnerabilities, for example, a newly discovered hardware or software defect enabling the creation or use of a new exploit. The process of identifying and analysing dynamic and emergent risks is examined further in Appendix D.1.2.

D.1.1 Identification and analysis of system and systemic risks

When addressing risks it is essential that the risk analysis, management and mitigation takes into account technology, process and people aspects in a holistic manner, and that the approach delivers a defence in depth rather than treating them as individual risks that can be managed independently. This holistic approach will also enable dependencies between differently affected elements, for example, a process and its underpinning or enabling technology, to be addressed in a consistent fashion. There are a number of tools available that support the management of risk in a coherent and integrated fashion. See Appendix I for further information on some of the available risk management guidance material.

These risks will fall into one of three groups:

(a) Technology risks – associated with:
   i. the architecture of the building systems; and
   ii. the integration, interfaces, infrastructure and implementation of building systems.

(b) Process risks – arising from the design and operation of any building systems related to:
   i. poor process design (for example, failure to implement a separation of duties);
   ii. failure to monitor and manage the operation of processes (i.e. to detect non-compliance or deviation from approved process and procedures); and
   iii. processes that are no longer fit for purpose, for example, the management of building access control lists and permissions.

(c) People risks – arising from:
   i. awareness and training issues;
   ii. failures to manage staff;
   iii. insufficient staffing levels;
   iv. loss or dilution of control through the inappropriate use of contractors, temporary staff or outsourcing; and
   v. intentional sabotage or disruption by disgruntled employee or contractor.

Technology-related risks may be identified using standard engineering and systems analysis techniques that are often applied to safety critical systems, for example, FMEA or AFD. The latter technique employs a method by which potential failures are identified not by asking what might go wrong, but by asking how someone can make it go wrong and how that failure could be prevented or mitigated. This encourages the generation of scenarios from combinations of single failures that might have greater impact than individual failures. Using these engineering techniques can help to both identify potential vulnerabilities and to explore the consequences of a number of risks that may interact at one time. The techniques can also be used to identify single points of failure within building systems and the overall building systems architecture, and to analyse faults that affect system and sub-system reliability and availability.

For existing processes, the related risks may be identified through internal audits, where failures to implement the designed processes and procedures may be detected, and process weaknesses identified. For new processes, or processes that are being modified, business process analysis and business process improvement techniques should be employed. From a cyber-security perspective, the objective of this processes analysis is to identify potential process steps where failure or incorrect actions could compromise the security or integrity of the building systems. For example, the operational effectiveness and integrity of a building's access control systems could be compromised through poor processes related to the management of lost access control tokens or passes, or the failure to manage the deactivation and recovery of these tokens and security passes from leavers. A consequence of either of these failures could result in unauthorised people gaining access to the building and/or to areas housing sensitive building systems.

The people-related risks relate to the effectiveness of the organisation's management of the behaviour of insiders, i.e. individuals with authorised access to building systems. These risks may be identified through activities such as internal audits, training needs analysis and reviews of job roles.

D.1.2 Identification and analysis of dynamic and emergent risks

Over the lifecycle of a building and its systems, threats and vulnerabilities will emerge that could not reasonably have been envisaged at the time that the building system was designed or implemented. Protecting building systems from these dynamic and emergent threats requires the ability to perceive threat activity and emerging vulnerabilities in the context of the building, its systems and associated processes, so as to permit active defence of the building systems and data.

As illustrated in Figure D.2, situational awareness involves gathering intelligence on what is happening in both cyberspace and the real world, so as to understand how information, events, and the organisation's own actions will impact cyber-security, both immediately and in the near future. Situational awareness enables the building systems' owner to identify possible, probable and real threats to building systems, by identifying and evaluating environmental changes so as to assess their potential impact. Examples of these changes include the emergence of new threat agents, publicity about a new vulnerability, evidence of reconnaissance from system logs or changes in personnel using the building systems.

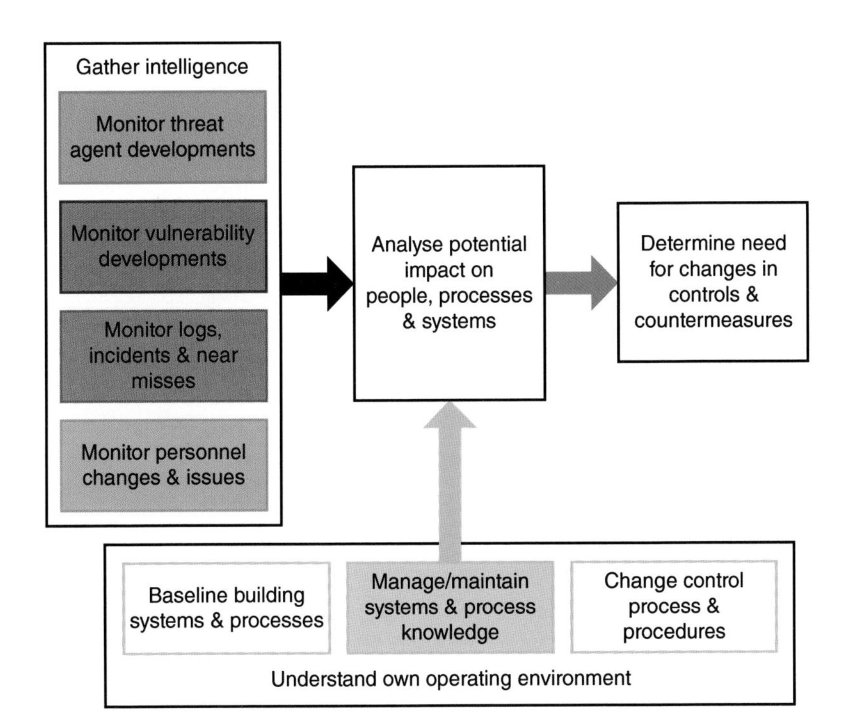

Figure D.2 – Situational awareness processes

In Figure D.2 there are four main processes, which are described in more detail below.

Gathering intelligence

This process involves identifying potential changes or developments that may have cyber-security implications for the building and its systems. The sources of intelligence will vary depending on the nature and use of the building, and may include open source material, information from vendors and systems integrators, warnings from national or sector specific response teams (for example CERTs or WARPs), advice and warnings from cyber-security specialists and, in the case of the critical national infrastructure, advice or warnings from national security services.

Intelligence gathering should cover the monitoring of:

(a) threat agent developments – this involves identifying potential new threat agents who may want to launch a cyber-attack on the building's systems, for example, the emergence of a new group of hacktivists who are planning major protests about the business activities of the building occupier.

(b) vulnerability developments – this requires the identification of security advisories that identify vulnerabilities or potential vulnerabilities in software and system components used in any of the building systems. The Heartbleed bug is a well-publicised example of a software defect in a particular release of the OpenSSL application that may be exploited to attack a system.

(c) Monitoring logs, incidents and near misses – the 'cyber kill chain' concept identifies a seven stage process involved in the more sophisticated cyber-attacks. The first stage of a cyber-attack is normally reconnaissance by the hostile party, involving target identification, information gathering and the evaluation of system and organisational structures. These activities may leave evidence in system logs and occasionally trigger system alarms. Regular reviews of system logs to identify anomalous or unusual events aids the detection of early attack stages. Monitoring of incidents and near misses can also help to identify potential vulnerabilities and enable suitable action to be taken.

(d) personnel changes and issues – indicators of potential insider threats take a number of forms, including changes in an employee's behaviour or employment status. For employees and contractors with access to building systems, particularly those who have administrative or elevated privileges, there are significant risks if they become disaffected. For example, if they have resigned and are working their notice, are subject to disciplinary measures, or have been given notice. The building systems' owner needs to be aware of such changes so that appropriate steps may be taken to monitor system access and use, and to revoke privileges or access should this become necessary.

Understanding the operating environment

Simply gathering the intelligence is of little practical value if its relevance and significance to the operation of the building and building systems is not understood. A lack of knowledge of the current operating environment can seriously hamper the investigation of whether an emerging threat or potential vulnerability may impact the building. Three activities need to be undertaken in order to develop and maintain an understanding of the operating environment:

(a) Manage/maintain systems and process knowledge – to understand the significance of the gathered intelligence, it is essential that the people analysing the intelligence have up-to-date knowledge of the building, its systems, data and the associated processes. These individuals will need access to current information on systems configuration, software and hardware versions, process descriptions and roles and responsibilities of the staff that perform the processes within the organisation.

(b) Identifying baseline information about building systems and processes – the availability of baseline information about building systems and processes will allow for intelligence evaluation and threat analysis. Inevitably, systems, processes and those responsible for managing and operating those systems and processes will change over time. The baseline therefore needs to be maintained under change and configuration control to ensure the availability of up-to-date information on the systems architecture, hardware and software inventories, system configuration, business processes and personnel roles and responsibilities.

(c) Create change control processes and procedures – these are necessary to ensure accountable and auditable management of the baseline information. The change control processes and procedures should be used to maintain the configuration, with records updated to reflect the implementation of approved changes. There will be circumstances where urgent changes need to be made to systems to maintain business continuity. In these circumstances, the changes may occur prior to receiving formal approval of the changes, but it is important that such changes are appropriately recorded so as to avoid a steady creep away from established baselines.

Analyse potential impact

Based on the gathered intelligence and up-to-date knowledge and understanding of the building's operating environment, the individuals responsible for the situational awareness analysis should use a systematic process to evaluate the intelligence. A triage process may be necessary to quickly filter out items of little or no relevance to the building, so that more effort can be applied to relevant material. The analysis should consider the impact of the threat or vulnerability on the building systems using the cyber-security attributes outlined in Appendix A.3. In addition to assessing the impact, the likelihood of a cyber-attack employing the threat or vulnerability should be assessed.

The analysis process can become quite complicated if impacts on technology, processes and people are being handled in a holistic manner. A number of tools have been developed to support risk analysis in complex environments. For all but the simplest situations, it is recommended that an appropriate risk management tool is used as a repository for the threat and vulnerability information and to provide analytic support as part of the risk management process.

Determine need for changes in controls and countermeasures

The final step in the situational awareness process is to determine whether any action is required. This will take into account the perceived impact of a cyber-attack, the likelihood of one occurring and the cost, disruption and complexity of implementing any proposed controls or countermeasures.

It is important to recognise that the decision on any particular threat or vulnerability could have a number of outcomes, including:

(a) no need for further action – the threat or vulnerability is not considered significant and in the case of a vulnerability, a cyber-attack exploiting it is thought to be very unlikely and of little consequence.

(b) decommission and removal of system – the threat or vulnerability affects a system that is no longer required or is delivering minimal benefits so, rather than providing additional controls or countermeasures, the system is decommissioned and removed.

(c) implementation of additional controls and countermeasures – the threat or vulnerability could have some impact on the building systems, so additional controls or countermeasures are appropriate and will be implemented as part of the organisation's routine business as usual change or maintenance activities.

(d) Immediate deployment of controls or countermeasures – the threat or vulnerability could have a serious impact on the building systems, so additional controls or countermeasures are urgently required and will be deployed immediately.

This process of reviewing controls and countermeasures should not be seen as a purely cumulative process, where layers of controls simply accumulate over time. The decision-making process should enable a continuous review of the appropriateness of all of deployed cyber-security measures. Systems upgrades coupled with research and development of new controls and/or countermeasures may enable implementation of new or more effective cyber-security solutions. It is also important that processes and procedural measures are both considered alongside technical controls. The latter may be easier to implement, but if they unduly constrain or hinder building operations, then occupiers and facilities staff will often find ways to bypass or circumvent them. The review process should also consider whether the defence-in-depth principle is being maintained or whether additional measures may be necessary.

D.2 Cyber security in the supply chain

Organisations procure a wide variety of products and services from a diverse range of suppliers, who will in turn procure products, components and services from other suppliers. The design, construction, operation and maintenance of a building can involve hundreds of suppliers with long international supply chains. This can create a number of significant cyber-security risks, particularly where the emphasis during procurement is to minimise cost, with the potential for quality and security requirements to be compromised.

To protect the building from cyber-security threats emanating from the supply chain, there is a need for increased visibility, traceability and transparency coupled with clear policy, processes and procedures to manage the procurement and delivery of products and services. There are likely to be requirements for the ongoing monitoring of suppliers, which will involve managing access to the building, building systems, associated processes and building data. These access-control arrangements need to cover both physical and virtual access. The suppliers' contracts need to include specific provisions addressing relevant cyber-security aspects.

Specific supply-chain related risks that need to be considered and addressed include:

(a) access to building systems – the arrangements for providing access and revoking access, when no longer required, whether on site or remotely via building related systems, for example, the building management system, access control system, fire alarm system, room booking and facilities management systems.

(b) access to building data – the nature of the access required will vary by supplier and role, but arrangements need to be addressed for access to BIM-related data and models and building occupant information (PII) used in the management of the building, etc.

(c) secure removal of building data that is no longer required – for example, the secure deletion of data from suppliers' IT equipment, cloud storage and backup, either when they cease to be a supplier to the building or when they no longer need to have access to the data, depending which occurs first.

(d) cyber hygiene of the supply of information and communications technologies – the need to ensure that genuine (non-counterfeit) components are used, that software and systems are delivered free from malware and spyware, that the default configuration disables guest or remote access accounts and that default passwords have been replaced.

(e) cyber hygiene of the use of personal devices, removable media, etc. – this is particularly relevant for suppliers' personnel that work in the building and may want or need to connect these items to building systems.

(f) delivery of trustworthy software – how is the safety, reliability, availability, resilience and security of software embodied in building systems assured and maintained.

An efficient way to handle these risks and the appropriate controls and countermeasures is to have a clear set of cyber-security policies, processes and procedures for suppliers, which are referenced in their contracts. Provision should be included in the contracts for these documents to be updated and reissued in the light of changes to the building, its systems and associated processes, and the cyber-security policy for the building.

Example: cyber security should not increase contract cost

A major engineering company undertook a review of equipment that was being delivered by its control system's suppliers to establish their compliance with industry sector recognised good practice. The suppliers were informed that they were required to comply with the published cyber-security good practice, for example – to avoid using hard-coded or default passwords, and that the engineering company was not prepared to pay extra for this compliance. With the exception of one supplier, all of the other control systems suppliers have agreed to meet this requirement and have been placed on a preferred supplier list. The supplier who was not prepared to meet the requirement is no longer a supplier to the company. This is significant because in many engineering environments, systems integrators charge their clients for the work required to achieve compliance with cyber-security good practice, and short-comings are often only picked up during penetration testing and may require further chargeable work to rectify.

D.3 System operations

D.3.1 Systems and infrastructure documentation

One cause of resilience and cyber-security failure is a lack of up-to-date documentation on infrastructure and systems design and configuration. This can lead to operator errors, damage to cables and compromised security and operational controls. A critical operational control for all buildings is the availability of up-to-date documentation for building systems, including hardware and software assets, interconnection and configuration information.

Whilst the use of BIM to develop an 'as built' repository of information about the building and its infrastructure may help to alleviate this problem in future, it will not address this information management problem for many existing buildings. Additionally, it will only be of practical use if the BIM data is maintained and updated to reflect post-occupancy changes to the design and configuration of the building and its infrastructure.

There will be a need to maintain good access control over the systems and infrastructure documentation. Failure to do so will allow a potential threat agent to conduct hostile reconnaissance and establish weaknesses of single points of failure in the building, its systems and the building-related infrastructure. It is also good practice for an up-to-date copy of the documentation to be held securely at a location outside of the building for use in a business continuity situation.

D.3.2 Building and systems maintenance

The way in which the building and building systems maintenance is undertaken can have a profound effect on the overall cyber-security protection of the building. Complex control systems require extensive maintenance across their lifecycle, including the replacement of failed system components, installation of patches and minor configuration changes to maintain service quality or functionality. Depending on the building's threat profile there may be a need for detailed policies and procedures relating to the maintenance of critical building systems. For example, where changes are required to access control systems, there should be an appropriate testing process implemented to ensure that the changes do not compromise the operation of the system.

Depending on the risk assessments undertaken whilst preparing or updating the building's cyber-security policy, appropriate cyber-security processes and procedures should be adopted for the maintenance of the building and its surroundings, the building systems and infrastructure. If maintenance activities are being undertaken by third parties, appropriate security briefings should be provided, along with copies of relevant policies, processes and procedures. The building owner or occupier should also put into place mechanisms that will ensure that the third parties adhere to these requirements.

For all but the simplest buildings, a building systems maintenance strategy should be established. This should comprise a set of policies, procedures and actions designed to support, promote and implement the following maintenance objectives:

(a) to keep building-systems related hardware in good working order;
(b) to keep building-systems related software, operating systems and system environments in good working order;
(c) to ensure that the status of the building-related systems meets existing organisational, industry and other accepted 'best practices' as they relate to operational and security requirements; and
(d) to maintain up-to-date configuration information, settings and documentation for the building, building systems and related infrastructure.

To implement the strategy an organisation will need to undertake three major categories of maintenance.

(a) preventive – the tasks performed on a system to correct or prevent degradation of performance, and to correct other minor issues prior to such degradation becoming larger problems.
(b) scheduled – this is a system of performing ongoing, routine maintenance procedures at periodic scheduled intervals. The purpose behind scheduling routine maintenance tasks such as upgrades, patches, cleaning and installs is to provide a measure of predictability and to move any expected downtime to off-peak hours.
(c) corrective – maintenance of last resort. The system is broken and must be repaired or replaced. Yesterday or sooner, will be the users' preference. Corrective maintenance can range from a simple component swap to replacement of an entire system. Corrective maintenance may be performed by in-house personnel, outside vendors or a combination.

D.4 Incident response, investigation and management

Incidents happen: cyber-security incidents are a daily occurrence, although many go unreported. As explained in Appendix D.1, there is a need to maintain situational awareness to enable risk management of emerging vulnerabilities and threats. This situational awareness should apply to both the enterprise and building systems and their associated business processes.

The impact of incidents on both the organisation and its reputation are often determined by the organisation's preparedness and its response to the unplanned event. The organisation's business continuity plan and the building's cyber-security policy should establish responsibility for developing and implementing an incident response infrastructure (for example, plans, defined roles, training, communications and managerial oversight). The effectiveness of this infrastructure will determine the speed of the response following the discovery of an attack and thereafter how effectively damage is contained, the attack terminated and normal operations restored.

Preparation is essential; when an incident occurs it is too late to start developing new procedures, considering reporting and data collection, and worrying about management responsibility, legal issues and the organisation's communications strategy. A cyber-attack can be as much of a business continuity issue as a major fire or flood, particularly if the incident causes a building to be harmed or makes it uninhabitable. Evidence from studies of cyber-security breaches suggests that there is often a significant period, usually measured in months, between an attacker gaining access to a system and the attack being detected. To minimise impact, the organisation needs to follow good practice, identify and contain the damage, remove the attacker(s) from the systems, and recover in a safe and secure fashion. In the event of a cyber-attack on a building or building system it is not just about the potential risk of exfiltration of sensitive data, it is also about restoring the trustworthiness of the systems. Consideration also needs to be given to the capture of any forensic data following a cyber-security incident. This is a particular issue with control systems, for example, building management systems, where the opportunity to collect evidence may be reduced due to a need to restore or maintain systems operations so that the building can continue to function.

Cyber security of buildings is not just about defending the building systems from hackers and hacktivists, it is also about the continuity of operating building systems in the face of adversity. This should include addressing the potential security and safety issues arising from natural events, for example, severe weather, earthquakes, solar storms, etc. Some of these incidents will have a direct impact on the building itself, others may affect, for example, the continuity and availability of mains power as illustrated by Hurricane Katrina and its impact on buildings in New York.

Key steps in preparing an effective incident response plan include:

(a) creating written incident response procedures, which will include the definition of both personnel roles for handling incidents and the phases of incident handling;
(b) assigning job titles and duties for handling building systems and related infrastructure incidents to specific individuals;
(c) defining which management personnel will support the incident handling process, identifying their key decision-making roles and their key contacts with the suppliers responsible for supporting the building, building systems and infrastructure;
(d) defining escalation procedures and identifying the circumstances under which regulatory and security or law enforcement bodies need to be notified.
(e) devising standards for the time required for facilities managers, building systems suppliers and support contractors to report anomalous events to the organisation's incident handling team. The mechanisms to be used for such reporting should be clearly identified, as should the kind of information that must be included as part of the incident notification.

(f) creating and maintaining up-to-date information on third-party contacts that may be required to assist in the event of a cyber-security incident. It may be prudent for hard copies of this information to be securely held by the building's security and facilities managers so that it is still accessible even if the building systems are inoperable.

(g) publishing information for all personnel using the building, including employees, agency staff and contractors, about the arrangements for reporting building systems anomalies and incidents to the organisation's incident handling team. This information should be included in the new joiners' pack and provided to building occupants as part of the routine security and business continuity awareness activities.

(h) conducting periodic incident management rehearsals, which include cyber-security related scenarios. These sessions should involve personnel and contractor representatives associated with the incident handling team and will help them to understand current threats and risks, as well as ensuring their familiarity with their roles and responsibilities in supporting the incident handling team. The testing or rehearsals should include all pre-defined communications, command and control processes, who it is that should deal with the issue, who it is that should manage the teams, and who it is that should manage stakeholders (for example, personnel, contractors, customers, press, security services, police/law enforcement, etc.). For major organisations the testing should also include a review of press releases and other pre-planned communications to verify that it is fit for purpose and conveys the right message in the event that a cyber-security event disrupts day-to-day operations.

To investigate security incidents, there may be a need for investigators to have access to items of personal IT equipment that have been brought into the building or connected to building systems. Provision should be made in employment and supplier contracts for digital forensic examination in the event of a cyber-security breach.

APPENDIX E

Configuration control

In the event that a threat agent is trying to compromise the building systems, the threat may be partly mitigated by a secure configuration of the networks and systems of both the organisation and the building. Specific control measures that can be used by the building owner and their facilities manager to prevent unauthorised access or actions include:

(a) change management for processes, systems and infrastructure;
(b) implementation of physical security measures to protect systems and infrastructure;
(c) maintenance of an inventory of authorised and unauthorised hardware and devices;
(d) maintenance of an inventory of authorised and unauthorised software;
(e) implementation of secure configurations for hardware and software;
(f) implementation of secure configurations for communications and network devices;
(g) application of wireless device control;
(h) limitation and control of communications, network ports, protocols, and services; and
(i) control of the use of administrative privileges.

E.1 Change management for processes, systems and infrastructure

A key part of configuration control is the operation of an effective change management process, where requests for change (RFCs) are subject to a formal assessment of threats and vulnerabilities before being approved for implementation. This allows for proactive management of new vulnerabilities, with appropriate countermeasures or remediation deployed as part of the change.

E.2 Implementation of physical security measures

Physical protection of the systems and infrastructure typically involves the protection of:

(a) cables routes from accidental or malicious damage;
(b) patch racks, etc. from unauthorised reconfiguration or changes; and
(c) hardware and devices from unauthorised tampering, interference, destruction or theft.

E.3 Maintenance of an inventory of hardware and devices

This involves actively managing the hardware inventory of components in the building systems and associated infrastructure. It will involve tracking items and correcting the records for all hardware and devices within these systems. It should also include the management of any networking or communications infrastructure so that only authorized devices are given access, and unauthorized and unmanaged devices are found and prevented from gaining access.

This managed control of all hardware and devices plays a critical role in the vulnerability analysis and in the planning and execution of systems backup and recovery. It also enables the responsible manager to ensure that these elements are appropriately maintained, including any critical firmware and security updates. It also allows the management and, where necessary, the isolation of any additional systems that may be connected to the building's network or infrastructure (for example, demonstration systems, temporary test systems, etc.).

E.4 Maintenance of an inventory of software

This involves actively managing the software inventory in the building systems and associated infrastructure. It will involve tracking software packages and versions (or releases) so that only authorized software is installed and can be executed, and that any unauthorized and unmanaged software is found and prevented from installation or execution. The management of software should include the management of patches and service releases.

As is the case for the maintenance of an inventory of hardware and devices, this managed control of all software plays a critical role in the vulnerability analysis and in the planning and execution of system backup and recovery. Organisations that do not have complete up-to-date software inventories typically have difficulty locating systems that are vulnerable or have been compromised with malicious software. This slows down the response to new vulnerability disclosures and the ability to mitigate incidents or to root out attackers.

Poor management of software configuration increases the probability that computer and other devices are either running software that is not required for business purposes, introducing potential security flaws, or running malware introduced by an attacker after a system is compromised. Once a single machine or device has been exploited, attackers may use it as a staging point for collecting sensitive information from the compromised system and performing reconnaissance on other connected systems. Attackers may quickly turn one compromised machine or device into many and traverse across network and organisational boundaries.

Example: use of a weak point in an enterprise's technical architecture to launch an attack

A serious attack on the point of sale system of Target Stores, a major US retailer, in Winter 2013 started with an attack on the HVAC supplier. By obtaining the login credentials for this supplier the attackers were able to log into Target's supplier portal. Through this system they managed to introduce malware into Target's IT infrastructure, perform hostile reconnaissance and then install malware on the company's point of sale system. This attack was aimed at theft of credit card data, but could easily have been an attack from an enterprise system on building systems with the aim of disrupting physical store operations.

E.5 Implementation of secure configurations for hardware and software

The aim of this activity is to establish, implement and actively manage the security configuration of all computer equipment, whether in the form of laptops, servers and workstations or hardware containing embedded software for processing and control purposes, for example, programmable logic controllers (PLCs) or remote terminal units (RTUs). This should involve the implementation of a rigorous configuration management and change control process in order to prevent attackers from exploiting vulnerable services and settings.

This is an important measure that needs to be carried out throughout a system's lifecycle. Typically, as delivered by manufacturers and resellers, default configurations of operating systems and applications are geared to ease-of-deployment and ease-of-use, i.e. they are insecure. It is common to find in their initial state, basic controls, open services and ports, default accounts or passwords, older (vulnerable) protocols and pre-installation of unneeded software. This makes it easy for an attacker to exploit the systems. This is a significant issue where legacy control systems are integrated with wider enterprise systems and thus inadvertently become connected to the internet.

The development of secure configuration settings is a complex task, typically beyond the ability of individual users. It may require potentially hundreds or thousands of options in order to make good choices. Even when a strong initial configuration has been developed and deployed, there will be a need to continually manage the security configuration so as to avoid security 'decay'. This decay may be a result of software updates or patches, particularly where suppliers reset options or settings, freshly reported security vulnerabilities and minor modifications to configurations to permit the installation of new software or support new operational requirements.

Example: insecure building management system

In June 2013, two security researchers reported that they could easily hack the building management system for the Google's Wharf 7 office in Sydney, Australia. The office used a building management system built on the Tridium Niagara AX platform, a platform that had been shown to have serious security vulnerabilities. Although Tridium had released a patch for the system, Google's control system had not been patched, which allowed the researchers to obtain the administrative password for it ('anyonesguess') and access control panels. The panels showed buttons marked 'active overrides', 'active alarms', 'alarm console', 'LAN Diagram', 'schedule' and a button marked 'BMS key' for Building Management System key.

Among the data the researchers accessed was a control panel showing blueprints of the floor and roof plans, as well as a clear view of water pipes snaked throughout the building and notations indicating the temperature of water in the pipes and the location of a kitchen leak.

Google confirmed the breach and advised that the company has since disconnected the control system from the internet. It is understood that the system could not be used to control electricity, elevators, door access or any other building automation.

For more information, see http://www.wired.com/2013/05/googles-control-system-hacked/

E.6 Secure configurations for communications and network devices

This activity relates to the configuration of communications and networking devices, for example, access points, GSM modems, and network devices such as firewalls, routers, and switches. The objective is to establish, implement and actively manage the security configuration of communications and network infrastructure devices. This should be achieved using a rigorous configuration management and change control process, with the aim of preventing attackers from exploiting vulnerable services and settings.

This is an important measure that needs to be carried out throughout a system's lifecycle. Typically, as delivered by manufacturers and resellers, default configurations of operating systems and applications are geared to ease-of-deployment and ease-of-use, i.e. they are insecure. The weaknesses that often need to be addressed include open services and ports, default accounts (including support or maintenance accounts), default passwords, support for older (and often vulnerable) protocols and pre-installation of unneeded software. This makes it easy for an attacker to exploit these system components. This is a significant issue where legacy control systems are integrated with wider enterprise systems and thus inadvertently become connected to the internet.

As with the previous activity – the securing of hardware and software configurations – this can be a complex task. However, it is important to limit the systems and connections that are exposed, as attackers search for vulnerable default settings, electronic holes in communications devices, firewalls, routers and switches and use those to penetrate system defences. The equipment should be set up to use the minimum set of services that meets the business and operational requirements, with unwanted features turned off or disabled. Again, it is important that this process is maintained across the system lifecycle to prevent security decay.

Example: insecure configuration vulnerability

In addition to problems caused by the poor configuration of security appliances, unpatched or unmitigated vulnerabilities can expose user systems. For example, Juniper Networks reported high severity vulnerability with Netgear DG834 ADSL Firewall Routers. This is an appliance that acts as an ADSL modem, a router, a 10/100 LAN switch and a Linux based IPTables firewall.

The appliance was reported to be prone to a firewall insecure configuration vulnerability, which affected the appliance when it is configured so that NAT (Network Address Translation) was disabled, causing the firewall to become ineffective. It is reported that this is because when NAT is disabled the main 'INPUT' IPTable entry is set to 'ACCEPT'. This may expose the appliance or internal hosts to malicious network traffic.

This vulnerability will result in a false sense of security where a user may believe that their network and appliance is protected when it is not.

For more information, see http://www.juniper.net/security/auto/vulnerabilities/vuln12447.html.

E.7 Application of wireless device control

This activity relates to policing the use of wireless devices on the building systems network and infrastructure. It involves deploying and implementing the processes and tools required to track, control, prevent, locate and correct the use of wireless devices, including wireless local area networks, access points, wireless client systems (for example, 3G, 4G and LTE modems), and the use of both commercial (for example, Bluetooth) and industrial wireless protocols as part of a system design.

These measures are necessary to prevent attackers using wireless access to bypass other security measures deployed by the organisations, for example, firewalls. Wireless attacks are a particular threat in urban areas or sites with a minimal secure perimeter outside the building. They represent a threat as they may be accessible from outside the physical building, bypassing organisations' security perimeters.

Wireless devices are a convenient vector for attackers to achieve and maintain long-term access into a target environment without the need for physical connections or a visible physical presence. The activity needs to include a search for, and detection of, unauthorized wireless access points on the networks, planted and sometimes hidden for unrestricted access to an internal network. These unauthorised devices may be installed by attackers or by insiders who are seeking more flexible use of their personal IT equipment within the building.

In addition to the threat of wireless devices being used as an attack vector to gain access to the building and its systems, there is also the issue of jamming, and interfering with, wireless signals. For example, Wi-Fi and Bluetooth signals are relatively easy to jam or disrupt and it is inadvisable to use these as the primary communications links for safety- or security-critical functions.

Example: insecure Wi-Fi leads to major breach at US retailer

In January 2007, the TJX Companies, which includes TK Maxx, announced that the firm had suffered an unauthorized intrusion into the 'computer systems that process and store information related to customer transactions'. The breach had allegedly occurred in December 2006.

In May 2007 it was reported that the hackers who stole 45 million customer records from the parent company of TK Maxx did so by breaking into the retail company's wireless LAN. The data theft, which included millions of credit card numbers, apparently occurred during the second half of 2005 and throughout 2006. It was also reported that the attack initially happened outside of a Marshalls discount store in St. Paul, Minnesota in July 2005 (Marshalls is owned by TJX Cos.). The insecure wireless network at the Marshalls discount clothing store may have allowed the attackers to gain a beachhead in retail giant TJX Companies' computer network. Apparently, while TJX's other systems were upgraded to Wi-Fi Protected Access (WPA), the Marshalls store's wireless network that connected credit-card processing hardware to the company's server was not.

The WEP was deemed insecure. It was originally demonstrated to be broken in 2001 and subsequent attacks by researchers have resulted in techniques allowing the WEP 104-bit encryption to be cracked in under a minute on an 802.11g network.

For further information, see http://www.zdnet.com/wi-fi-hack-caused-tk-maxx-security-breach-3039286991/

E.8 Limitation and control of network ports, protocols, and services

A consequence of the increasing use of commercial off-the-shelf (COTS) products and software as components in industrial control and building systems is that there may be unnecessary software components installed and enabled by default. For example, if a component includes server software components, TCP/IP ports 20 and 21 may be enabled allowing FTP data transfer and control respectively. If an FTP service is not required then these ports and the corresponding service should be disabled.

This activity therefore relates to the ongoing management of the operational use of ports, protocols and services on networked devices in order to minimize available vulnerabilities. This activity is an essential step in the reduction of risks arising from commonly exploitable vulnerabilities, such as poorly configured web servers, mail servers, file and print services, and domain name system (DNS) servers. This management activity is often complicated by the failure of the software installation package to inform a user or administrator that the services have been installed and enabled. Attackers may attempt to exploit these services, often succeeding through the use of default user identities and passwords or other published exploits.

Example: managing default services

Tablet devices such as the Apple iPad are becoming increasingly common in the corporate environment, where they are used to access corporate email systems, files, etc. Their use is increasing in support roles such as facilities and system management where the device can be used to hold or provide access to support documentation. These devices provide both Wi-Fi and Bluetooth interfaces. In a corporate environment the latter may not be required and instead may be turned off in the device's control panel.

When an iOS upgrade, or patch, is installed the Apple software turns on the Bluetooth service by default. The user will then need to manually turn this service off if the risks associated with Bluejacking and Bluesnarfing are to be avoided.

E.9 Control of the use of administrative privileges

The misuse of administrative privileges is a primary method for attackers, whether internal or external, to attack systems and the organisation and to bypass many of the normal system security controls. Depending on what privileges the compromised account has, the attacker may also be able to manipulate the security and configuration of other systems, for example, creating unauthorised accounts or manipulating privileges to allow greater access by other compromised accounts.

This activity relates to the processes and tools that should be put in place and used to monitor, control and prevent or correct the use, assignment and/or configuration of administrative privileges on computers, networks and applications, both those affecting building systems and those in the wider enterprise architecture.

A common attacker technique is to take advantage of uncontrolled or poorly controlled administrative privileges. This may involve targeting a privileged user such as a system or IT manager with the aim of getting them to run malware that allows compromise of their account. Typically the malware will be used to install further attack tools, giving access to the user's machine and the connected networks. The success of this type of attack often depends on the user having elevated privileges, thus allowing the silent installation of the malicious software. Less privileged accounts may be unable to run this type of exploit.

Example: misuse of administrative privileges

The United States government has been severely embarrassed by disclosures leaked to the press by Edward Snowden. He was a contractor working as a systems administrator with access to sensitive government systems. Based on the limited publicly available information on how this breach occurred, it appears that Snowden executed his theft of data by:

(a) researching the target – he used his valid systems access to determine what information was available and where it was stored.

(b) initial intrusion – he used his position of trust to gain unauthorized access to other administrative accounts, some of which allowed him to gain full, trusted status to information he was not authorized to access. Having 'root' or equivalent administrative status on the systems gave Snowden total access to all data. Just like an advanced and persistent external attacker, he took care not to set off alarms and he covered his tracks.

(c) exfiltration – to get data off the secure network, he could not simply save it to a USB or flash drive. Instead, data needed to be stealthily moved across networks to avoid detection. Just like common cybercriminals, he apparently used 'Command and Control' servers to receive encrypted data sessions.

A failure to manage administrative privileges and monitor the activities of trusted individuals allowed Snowden to steal a significant amount of highly classified information.

This breach is not uncommon; individuals with systems administrator or other elevated privileges are regularly reported as having been the cause of security breaches, data theft and fraud.

APPENDIX F

Managing 'people' aspects

A building is affected by two groups of people:

(a) Insiders, which include –
  i. those individuals who have access to the building, its systems, associated processes and data, either through employment by the building owner or occupier, or who provide services to the building and its occupants, for example, contractors, suppliers, etc.; and
  ii. visitors, i.e. those individuals who have access to the building on a temporary basis and who are not employed by the building owner or occupier, for example, members of the public visiting a government or local government building, customers using leisure facilities and guests at a business meeting or event.

(b) outsiders – those individuals who do not have authorised physical or virtual access to the building, its systems or the surrounding area, for example, deliberate trespass by protestors or criminals, unauthorised remote access to building systems including any non-public Wi-Fi, etc.

Cyber-security measures need to address the threats posed by both groups and recognise the differing levels of control and sanctions available to enforcing the building's cyber-security policy. This section addresses the handling of the first group.

F.1 Appointments, roles and responsibilities

F.1.1 The insider risk

Insiders are a major source of risk and, as explained in Appendix A.5, their actions may or may not be malicious. This section considers some of the steps that can be taken to reduce the risk of a cyber-security incident caused by an insider. CPNI conducted a study in 2013, which established that there are five main types of insider activity:

(a) unauthorised disclosure of sensitive information (either to a third party or to the media);
(b) process corruption (defined as illegitimately altering an internal process or system to achieve a specific, non-authorised objective);
(c) facilitation of third party access to an organisation's assets (including premises, information and people);
(d) physical sabotage; and
(e) electronic or IT sabotage.

The most frequent types of insider activity identified were unauthorised disclosure of sensitive information (47%) and process corruption (42%). From a built environment perspective, examples of these activities could include:

(a) unauthorised disclosure of sensitive building plans relating to building security measures, or PII of employees and visitors;
(b) corruption of access control and pass issuing processes to allow escalation of access privileges or to remove evidence of unauthorised access to sensitive areas;
(c) sabotage of building systems by physical or electronic means; and
(d) manipulation of other employees or contractors (i.e. deliberate attempts to acquire information or access by manipulating staff).

Almost all cyber-attacks can be assisted or conducted by an insider, whether motivated by criminals, terrorists or competitors seeking a business advantage. This could be anyone with authorised access to the building, its systems, the associated processes or building data, including employees or any contract or agency staff (for example, cleaners, caterers, security guards). An insider may already be working for the building owner or occupier, or may have recently joined so as to infiltrate the organisation specifically to exploit the access that the job or role might provide.

Example: disclosure of sensitive building-related information

In January 2008, a former agency worker was found guilty along with five others for the £53 million robbery of the Securitas depot in Kent in 2006. The agency worker based at the depot was on a low wage, but worked in the cash handling operation and had extensive access to a facility that dealt with hundreds of millions of pounds in cash. He was accused of providing information to the criminal network ahead of the raid, using a hidden camera to film the inside of the depot.

For further information, see http://news.bbc.co.uk/1/hi/england/7214598.stm and http://news.bbc.co.uk/1/hi/uk/7154191.stm

Example: attempted sabotage of building infrastructure

In August 1999 a former security guard who had worked at the stadium of Charlton Athletic was found guilty of a conspiracy to cause a public nuisance by plotting to sabotage the floodlighting at the stadium during a Liverpool v Charlton Athletic fixture. The scam was discovered when two Malaysians and a UK businessman were caught with a circuit-breaker at the stadium three days before the match. They had planned to plant the electrical device to sabotage the floodlighting. It was to be triggered with a remote control unit when the score favoured a betting syndicate. The potential for huge profits meant that they could promise to pay the security guard £20,000 to let them into the ground to plant the device.

For further information, see http://news.bbc.co.uk/1/hi/uk/426092.stm

In addition to the malicious insider threat there is always the risk of non-malicious insider actions. These threats typically arise from factors such as errors, omissions, ignorance or negligence. Social engineering attacks on an organisation tend to rely on naïve behaviour by insiders, who may fail to follow security processes and procedures. Examples of insider behaviour that can be exploited by cyber-attackers include:

(a) compromising physical security measures by allowing tail-gating into sensitive or restricted areas or leaving security doors and fire exits propped open;
(b) gratuitous disclosure of information about building systems on social media sites, for example, information about using specific systems on LinkedIn or posting pictures on Twitter or Instagram that aid hostile reconnaissance;
(c) disclosing account security information in response to a phishing or other social engineering attack; and
(d) careless use of email, which results in sensitive information being sent to incorrect addressees.

F.1.2 Factors contributing to insider attacks

There is a clear link between insider acts taking place and the presence of exploitable weaknesses in an organisation's protective security and management processes. Factors occurring at the organisational level include:

(a) poor management practices;
(b) poor use of security and systems auditing functions;
(c) lack of protective security controls;
(d) poor security culture;
(e) lack of adequate, role-based, personnel security risk assessment;
(f) poor pre-employment screening[7];
(g) poor communication between business areas;
(h) lack of security awareness of personnel risk at a senior level; and
(i) inadequate corporate governance.

F.1.3 Managing personnel security

To protect the building from insider-related cyber-security threats it is important to put into place an effective and holistic scheme to manage personnel security. An example of an effective framework is the Holistic Management of Employee Risk (HoMER) developed by CPNI[8]. This framework aims to reduce the risk of employees' behaviour damaging an organisation. The term 'employee risk' is defined as counter-productive behaviour, whether inadvertent, negligent or malicious, that can cause harm to the organisation.

The HoMER guidance leads an organisation through the key stages of a people risk-management lifecycle:

(a) vision and leadership – solid and engaged leadership, corporate governance and transparent policies in managing people risk and strengthening compliance;
(b) assess – adoption of a demonstrable risk-based approach and the implementation of reliable asset, access and identity management;
(c) protect – developing compliant policies and procedures for protective monitoring;
(d) respond – preparation in advance for handling employee incidents, including guidance on the best way to manage an incident to minimise damage and maintain stakeholder trust;
(e) recover – effective steps for post-incident recovery, including processes for maximising lessons learned to improve security.

---

[7] See for example BS 7858:2006, the Code of Practice for Security screening of individuals employed in a security environment

[8] CPNI (2012) – HoMER Guidance. http://www.cpni.gov.uk/advice/Personnel-security1/homer/

The principles set out in the HoMER guidance can be used to reduce the personnel risk to the building if they are applied to the facilities management and IT teams responsible for the management of the building, its systems, associated processes and building data. Whilst the HoMER guidance was developed for use with employees, its principles could be applied to all contractor, sub-contractor and agency personnel who are normally working in the building and its surroundings.

F.2 Managing consultants, contractors and agency staff

Many organisations make extensive use of consultants, contractors and temporary or agency staff throughout a building's lifecycle. This ever-changing group of insiders who often have privileged access to the building, its systems and data can pose a serious threat to the safety and security of the building. They will, for example, often know considerably more about the vulnerabilities in the building and its systems than the owner or occupier. During periods of change they may have access to significant volumes of sensitive information, for example, about the organisation's strategy, plans, etc. Facilities management teams are often involved from the very early stages in planning special events, and are likely to know about the presence of VIPs and dignitaries and any special security measures being put into place to protect them.

The use of contractors and agency personnel can result in an increased personnel security risk, as the individuals' loyalty may not be to the organisation that engages them, and they may have little or no regard for the security culture or needs of the organisation. This is potentially a significant issue in the architecture, construction, engineering, facilities management and IT industries, where the individual undertaking the work may have no direct contractual relationship with the organisation and may operate at the end of a long chain of sub-contractors. These levels of sub-contracting increase the risk that the security standards applying to the organisation and building become confused, diluted or even simply ignored.

To manage the insider threat posed by building- or facilities-related contractors and agency personnel, organisations need to implement the following measures, as illustrated in Figure F.1, as part of their contracting process:

(a) risk assessment – to identify and assess the insider risk arising from the use of contractors and agency personnel in specific roles or for specific duties;
(b) pre-engagement screening – to ensure that only trustworthy and competent contractors and agency personnel are engaged;
(c) communicating security requirements to contractors – this should address the required pre-engagement screening as well as the personnel, physical and cyber-security measures that will apply to the contract, any sub-contracts and to all contractor or agency personnel;
(d) embed ongoing personnel security into contracts and practice – ensure that contracts include appropriate measures and controls and establish the practices required to minimise opportunities for contractors and agency personnel to abuse the organisation's assets, including building systems, associated processes and building data once engaged;
(e) ensure that security requirements, procedures and any required pre-employment screening are cascaded throughout the entire contracting chain;
(f) ensure that there is contractual clarity about any responsibility for damage from security lapses or breaches, including the right to approve the choice

of sub-contractors and to require contract termination in the event of poor security performance; and

(g) establish and implement audit procedures of contractors – use periodic audits to encourage compliance with security policies, processes and procedures throughout the contracting chain.

It is important to recognise that the risk from contractors and agency staff is not confined to those who work in, or who have regular physical access to, the building. It is becoming increasingly common for technical support to building and control systems to be provided in part through remote connections by the service engineers and technicians. These largely invisible individuals may have considerable control over these systems and, due to the nature of their work, may be subject to minimal supervision by the organisation's own personnel.

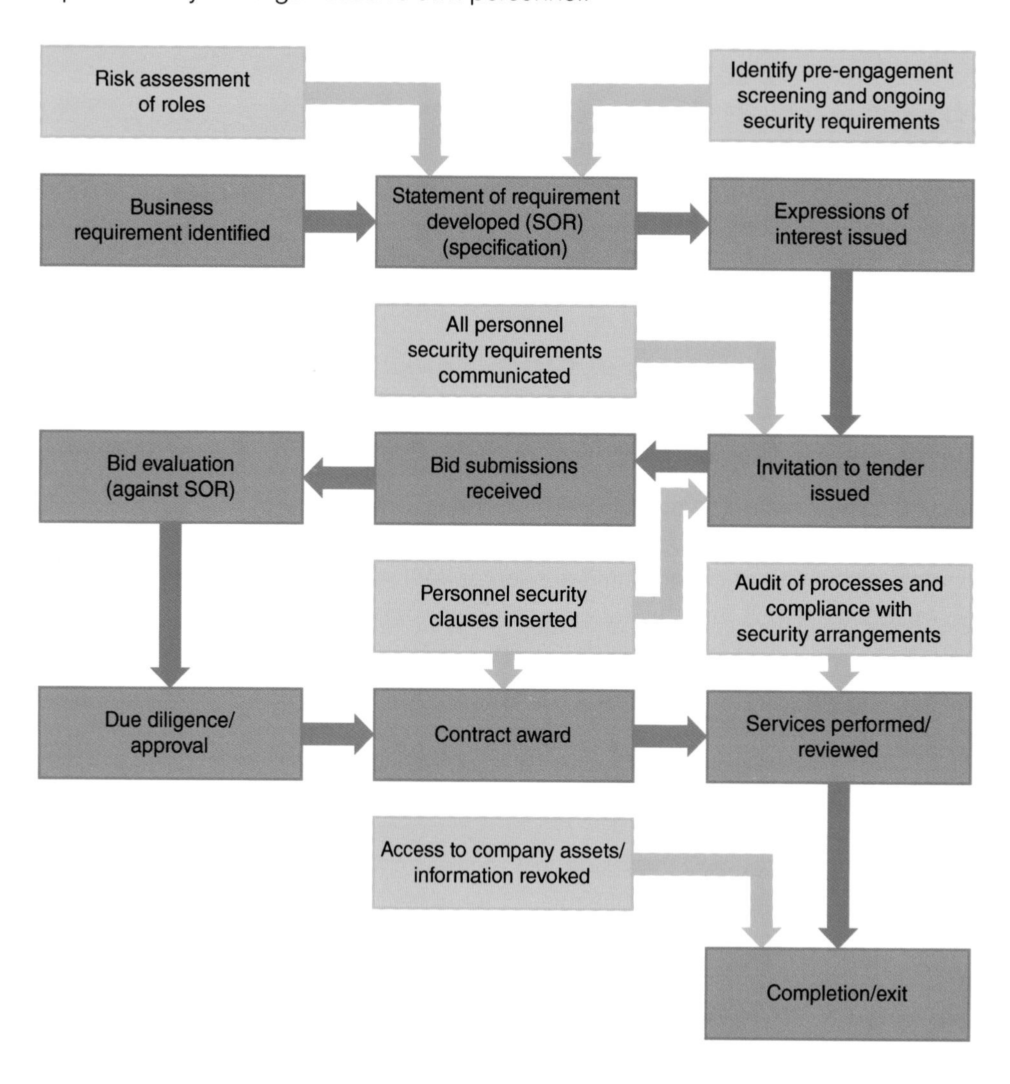

Figure F.1 – Personnel security in the contract lifecycle[9]

Whilst the steps described above and illustrated in Figure F.1 address the needs for the contracting process, the behaviour of the individuals needs to be addressed. There are a number of steps that should be taken with the individuals:

(a) include a security briefing as part of the contract induction required for non-employees who work at the site/building, or who have access to the systems;

(b) include them in any regular awareness and security training;

9 CPNI (2011) The secure procurement of contracting staff – Figure 1

(c) ensure that they are aware of any acceptable use policies and cyber-security policies if they have access to the building's or the organisation's systems;

(d) ensure that they understand any rules relating to the handling and, where necessary, the destruction of sensitive information;

(e) ensure that they understand site/building security procedures, including rules relating to the handling of removable media and personal IT equipment; and

(f) ensure that there are effective exit procedures to be applied when an individual ceases to work on the contract. These should include revoking access to the site/building, return of passes, keys, remote access tokens, equipment and contract- or building-specific clothing or personal protective equipment, removal of any building or organisation data from any removable storage or personal IT equipment used during the contract, return of documents (physical or electronic) and removal of access to building systems (including any remote access).

Example: managing remote access to control systems

In November 2011 there were press reports that hackers had gained remote access to the control system of the city water utility in Springfield, Illinois. It was suggested that their actions had destroyed a pump. The allegations arose when a water district employee noticed problems in the city's Supervisory Control and Data Acquisition System (SCADA). The system kept turning on and off, resulting in the burnout of a water pump. It was suggested that hackers may have been in the system as early as September 2011. The press reports suggested that intruders launched their attack from IP addresses based in Russia and gained access by first hacking into the network of a software vendor that makes the SCADA system used by the utility. The hackers then stole usernames and passwords that the vendor maintained for its customers, and used those credentials to gain remote access to the utility's network.

However, within a week these reports were contradicted by the US DHS, who could find no evidence that a hack had occurred. In truth, the water pump simply burned out, as pumps are wont to do. The presence of a Russian IP address in the SCADA log was the result of a call to the contractor, who helped set up the utility's control system and provided occasional support to the district. In June, the contractor and his family were on vacation in Russia when the water utility called his cell phone seeking advice on a matter and asked him to remotely examine some data-history charts stored on the SCADA computer. He did not mention he was on vacation in Russia, and simply used his credentials to remotely log in to the system and check the data.

Whilst this incident turned out to be legitimate access by a contractor, it highlights the need for appropriate procedures for remote access, both in terms of who can access it and from where it can be accessed.

For further information, see http://www.wired.com/2011/11/hackers-destroy-water-pump/ and http://www.wired.com/2011/11/water-pump-hack-mystery-solved/

F.3 Awareness, training and education

A successful and effective cyber-security programme for a building comprises three distinct components:

(a) development of a cyber-security policy that reflects business needs, based on known risks to the building, its systems and associated processes, and the building data;
(b) informing building users and building-related contractors of their cyber-security responsibilities, as documented in the cyber-security policy, processes and procedures for the building; and
(c) establishing processes for monitoring and reviewing the programme.

The development of the policy, processes and procedures was described in Appendix C, and the monitoring and review should be part of the cyber-security strategy described in Appendix B. The arrangements for informing users and contractors of the building's cyber-security policy fall within the personnel security arrangements described in Appendix F and, in particular, the provision of cyber-security awareness, training and education described in this section.

Cyber-security awareness, training and education are a progression or continuum as illustrated in Figure F.2. The development of an awareness and training programme should be linked to a training needs analysis, which in turn will relate to the building's cyber-security strategy, policies, processes and procedures. It may be possible to use and/or adapt awareness and training material available to the building owner, operator or occupier organisations to provide suitable material that includes and addresses the building systems.

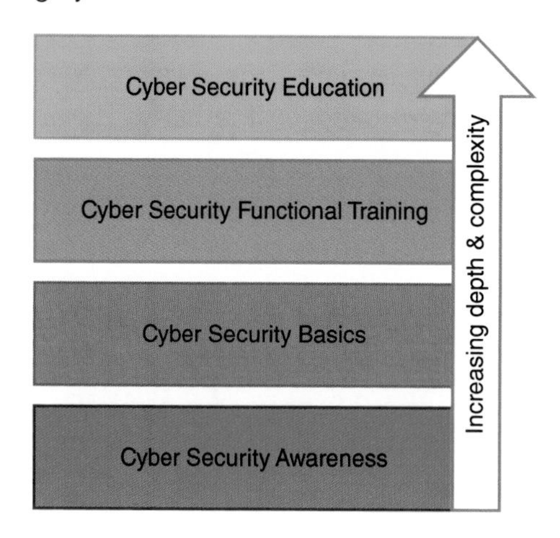

Figure F.2 – Progression through awareness, training and education

F.3.1 Cyber-security awareness

The lowest level in Figure F.2 is 'awareness'; this is not related to training. The objective of an awareness presentation is simply to focus attention on cyber security. Following an awareness presentation, individuals should be able to recognize cyber-security concerns and act or behave appropriately.

A significant number of topics could be covered in any awareness session. From a building cyber-security perspective, topics may include:

(a) physical security – access control, visitor badges, parking and deliveries, bomb threats;
(b) password use and management – creation, frequency of changes and protection;

(c) protection from malware – scanning of removable media, updating definitions;
(d) web and email use – allowed versus prohibited; monitoring of user activity;
(e) social engineering, phishing, spear phishing and handling of unwanted emails/attachments;
(f) incident response – who to contact and what to do;
(g) changes in the system environment – increases in risks to systems and data (for example, water, fire, dust or dirt, physical access);
(h) inventory and property transfer – identify responsible organisation and user responsibilities (for example, media sanitisation);
(i) handheld device security issues – address both physical and wireless security issues;
(j) access control issues – address least privilege (i.e. minimise use and granting of systems administration privileges) and separation of duties (i.e. sensitive tasks should require approval from two individuals);
(k) individual accountability – explain what this means in the organisation; and
(l) security-minded communications – reducing the amount of sensitive or potentially sensitive information that leaks out of an organisation through a variety of channels, including marketing and recruitment material.

Awareness presentations should be given as part of the site/building induction session for all personnel working in the facilities management and/or building management teams and their associated contractors and agency staff. Refresher training should be delivered periodically to update individuals on changes and reinforce their overall cyber-security awareness.

The 'cyber-security basics' level shown in Figure F.2 is effectively an extension of the awareness presentation material. It will typically address a topic in slightly more detail and will act as a bridge to the formal security training. For example, the basics training on passwords might provide more detailed advice on how to create passwords that conform to the organisation's policy and/or the use of 2-Factor Authentication for protection of more sensitive systems or as part of a remote access cyber-security regime.

F.3.2 Cyber-security training

Cyber-security training strives to produce relevant and needed security skills and competencies that will be used by functional practitioners outside of the cyber-security area, for example, operations and maintenance personnel being taught specific systems administration skills. Aside from the depth of coverage of a topic, the principal difference between awareness and training is that the latter is intended to teach specific skills that are required for an individual's job role, whereas awareness focuses an individual's attention on a specific issue or set of issues. Training courses may be delivered at various levels, for example, basic, intermediate and advanced.

Training needs will vary from building to building, but might typically include:

(a) systems administration for building-related systems;
(b) development of cyber-security policies, processes and procedures;
(c) risk assessment and management;
(d) maintenance and configuration of physical and cyber-security systems;
(e) personnel security;
(f) business continuity and contingency planning; and
(g) incident management and investigation.

Training will generally not lead to any formal qualifications but may provide credits towards, or some exemptions from, a formal educational course or be used in support of an individual's continuing professional development.

F.3.3 Cyber-security education

Cyber-security education aims to integrate all of the security skills and competencies of the various functional specialties into a common body of knowledge so as to produce cyber-security specialists and professionals with the knowledge and understanding to provide a proactive response to new or novel situations. A cyber-security education programme will generally be offered by a university or college and may result in the award of a certificate or degree depending on the nature of the course. A key difference between training and education is that education should provide the student with an understanding of the underpinning knowledge and principles behind the techniques they are taught as part of a training course.

F.4 Audits

Evidence of insider incidents show that two common factors are poor management practices and the poor use of auditing functions. Given the important roles that personnel security has in both the overall security and the cyber security of an organisation and building, there is a need for appropriate quality assurance measures to ensure compliance with policies. From a personnel security perspective this should cover:

(a) pre-employment screening of employees, agency staff and other temporary workers;
(b) pre-engagement screening of contractors;
(c) compliance with, and implementation of, ongoing security requirements, for example, the handling of disciplinary or other personnel-related actions following investigation of security breaches or incidents; and
(d) monitoring of security awareness training.

The right to audit must be specified in contractual documentation, which should also address any termination and compensatory arrangements in the event that an organisation or individual is in breach of their security obligations.

The organisation should have a clear policy, processes and procedures for the handling of audits. The processes should be transparent and, wherever practical, conducted by an independent party. The terms of reference and scope of any audits should be agreed in advance and, with the exception of situations where urgent action is required to investigate a security breach, reasonable notice should be given. Audits should address both the effectiveness of the processes and procedures in use and adopt a sampling approach to ensure that processes were completed to the required standard.

Extensive use of contractors during the design, construction and operation of a building increases the risk that the cyber-security policies relating to the building become diluted or confused as obligations are passed down the supply chain. To mitigate this risk, the top level contracts, i.e. those between the building owner (and any tenants) and principal contractors, should explicitly address:

(a) the cyber-security controls required, both pre-engagement and on an ongoing basis, of the contractor and that they are cascaded and upheld throughout the entire contracting chain;

(b) who is responsible for any lapses of cyber security and what actions or measures must be taken in the event of a lapse or breach;

(c) the right of the building owner and/or tenant to approve any subsequent choice of contractor; and

(d) the right of the building owner and/or tenant to audit the implementation of the cyber-security standards at any point in the contracting chain and, if issues are identified, to require them to be addressed by the relevant contractors.

Managing technical aspects

G.1 Operational security

The objective of operational security is to ensure the correct and secure operation of the building systems, including the maintenance of building data. Operational security covers a range of activities, including:

(a) equipment maintenance – ensuring that up-to-date systems architecture information is available for building systems and that equipment, plant and machinery is correctly maintained so that it meets the required availability and performance standards. This encompasses not only the information technology components of the building systems but also electro-mechanical elements, batteries, etc.

(b) asset management – maintaining accurate records of the equipment, configuration and software deployed in building systems. The asset management procedures should also include the bringing of equipment and software on site and the taking off site of equipment and software, for example, updating asset records, checking for malware, removal of sensitive data, managing the secure storage and the acceptance testing of spare equipment or components, etc. The maintenance of accurate change and configuration management records, for systems, architecture, hardware and software can support the identification and/or validation of potential vulnerabilities and investigation of system outages and security incidents.

(c) an arrangement for the secure disposal or reuse of equipment – to ensure that where equipment contains storage media and/or system configuration information, this has been securely removed before the equipment is removed from the building.

(d) protection of unattended user and control equipment – ensuring adequate protection is in place to prevent unauthorised use or access to control workstations and control panels.

(e) operational procedures and responsibilities – there should be properly documented operating procedures for building systems that cover all aspects of the operational activities associated with these systems. These should include procedures for the start-up and shut-down of systems, backup and recovery, maintenance, media handling, management of system logs, cyber-security incident management, induction, training and safety.

(f) business continuity planning and management – including maintenance and testing of business continuity policies, procedures and plans to ensure that they are fit for purpose and that relevant personnel are aware of their roles and responsibilities when responding to a business continuity event.

(g) change management – the formal procedures and processes that are applied when managing changes that affect the building systems and building data. Relevant changes may include those to the organisation, business processes, information or data processing and/or storage, and systems (design, configuration, interconnections, etc.). The change management processes should ensure that the impact of all aspects of a change that affects the building systems has been considered. An important part of this management process is the assessment of impact on the risk profile for the building systems, i.e. does the change introduce new vulnerabilities, expose the system to new threats, or reduce the exposure to existing vulnerabilities and threats.

As part of the equipment maintenance regime for building systems it is important that system capacity requirements are regularly reviewed. An active capacity management regime should ensure that the use of system resources is monitored and tuned and that projections of future capacity requirements are undertaken to maintain system availability and performance. The capacity management needs to apply to both IT systems and physical components. Examples of managing capacity demand include:

(a) maintaining disk space by removing obsolete data and moving equipment/system logs of the live system into archival storage as required by the operating procedures;
(b) decommissioning and removing unwanted software, systems, databases, environments and infrastructure, thereby reducing the risk of these items being used to compromise the building systems;
(c) managing the use of network and communications bandwidth to ensure that non-essential services do not restrict or compromise control and information flows for business critical systems; and
(d) monitoring and managing power load in relation to the capacity of power distribution systems, uninterruptible power supplies (UPS) and standby generators – including the fuel supply for standby generators to ensure availability in the event that they are required for a prolonged period.

As part of the maintenance and operating regime, business continuity arrangements need to be tested. Whilst it is common practice for business continuity plans to be tested by desk exercises, where individuals fulfil their allotted roles, there is also a need to test the recovery of systems. For complex buildings, which are heavily reliant on their building systems, there is a need for periodic testing of the building's 'black start' capability, i.e. all of the building systems have been shut down and need to be restarted. Depending on the local environment, this testing may be conducted assuming that there is no power available from the electricity grid. By performing this type of recovery testing once or twice a year, the building management team ensure that systems are recoverable in the event of a full outage. This testing can be linked to any periodic retesting/certification of systems and the building infrastructure from an electrical safety perspective.

G.2 Physical security of building systems

The physical security of building systems is a critical factor in maintaining their cyber-security and both need to be managed in a unified manner. Whilst the use of IP-enabled cyber–physical systems – for example, heating/ventilation/air conditioning (HVAC), lights, CCTV, access control – offers considerable benefits, it also opens up new vulnerabilities targeted for exploitation by hackers trying to access an organisation's network to steal information or impact corporate operations.

In many organisations one team oversees the physical operations of a facility, such as premises security and maintenance of building systems, but a separate team manages the organisation's IT operations and security of the corporate network. Yet there may be many instances where the network and communications infrastructure causes functions of both teams to overlap. In a worst case scenario, these teams are in separate departments, their personnel don't communicate on a regular basis, and the bulk of the work is outsourced to a variety of different contractors, with no contractual obligation for them to work together or cooperate.

Physical threats can and do have an impact on IT operations, just as cyber-security threats can have an impact on physical operations. Both threats therefore need to be managed in a unified manner, with the physical protection of building systems being accorded the same level of protection as the building's operational or user space. Security perimeters should be defined and implemented to protect building systems, their cabling and any associated plant and machinery. These security perimeters should be designed to prevent unauthorised access or tampering and, depending on the location and criticality, may need to be alarmed and monitored by CCTV systems. When considering the level and type of protection to be provided, a defence in depth approach is more reliable than a single protective barrier.

Example: protecting critical plant and machinery

Chillers forming part of an HVAC system that delivers a conditioned environment for equipment or a server room are often installed at ground level or on the walls of a building. Damage to, or interference with, these items can cause a malfunction of the HVAC system leading to over-heating of the room and shut down or damage to the equipment or servers. If these systems are critical to the building's or organisation's operation they should be located in a secure area with suitable physical barriers or relocated above ground level, for example, onto a roof or a wall above ground-floor level.

Examples of physical items located outside of a building that may need protection to maintain the cyber security of the building and its systems include:

(a) plant and machinery – chillers, transformers, switchgear, standby generators, fuel tanks (for the generators);
(b) cable and utility routes – for the supply of power, water, gas and telecommunications services. For campus situations this should include the protection of inter-building cabling used by telecommunications and networking systems; and
(c) CCTV cameras, access control devices – to prevent damage or tampering, and to prevent both physical and electronic attacks on the cabling that connects the cameras or devices to their monitoring and control systems.

Within a building, physical protection typically needs to be afforded to the following items to maintain the cyber security of the building systems:

(a) rooms housing building systems' control workstations;
(b) plant and equipment rooms;
(c) wiring distribution rooms, patch panels, cableways, ducts, etc.; and
(d) storage space for building systems' equipment when not installed and in operational use, i.e.
    i. on receipt, whilst awaiting acceptance and testing;
    ii. prior to use, for example, spares; and
    iii. while items await secure disposal.

The physical protection requirements identified above include the control of access and protection from damage (for example, from fire, flood, earthquake, explosion, civil unrest, etc.). The risk of damage should be taken into account when siting critical components both inside and outside the building. For example, if a building is potentially at risk from flooding, consideration should be given to locating sensitive systems above ground-floor level, both within the building and where items are installed outside.

Example: disruption to building systems and operations through flooding

On 24 December 2013 there was severe disruption to the operations of the North Terminal at Gatwick Airport following a prolonged period of very heavy rain. At around 05:00 flooding caused electrical disruption, which affected the North Terminal, resulting in most of the electrical and electronic services failing. The North Terminal was served by six switch rooms, only two of which had failed. However, although some systems remained operational, the loss of power led to system issues, which meant that:

(a) only 2 of 9 international baggage reclaim belts were operating and no power and no lights in the baggage reclaim hall;
(b) no check-in system, flight information screens, or telephone systems were available;
(c) only out-of-gauge luggage check-in belts were operating;
(d) toilet flushing mechanisms on the ground and first floors, which are driven by electronic systems, were not operating; and
(e) the passenger security screening facilities were also significantly affected, reducing the number of lanes available.

Investigations after the event found that the surface water gained access through the electric cable conduit system. This had not previously been identified by the airport's flood risk assessments as a source of flood risk.

For further information, see *Disruption at Gatwick Airport – Christmas Eve 2013*, report to the Board of Gatwick Airport Limited, available for download from: https://www.gatwickairport.com

G.3 Communications security, EMC and jamming

G.3.1 Communications security (COMSEC)

In addition to networks carried over fixed cabling, building systems use a variety of technologies to connect and control components. These range from simple serial communications protocols (RS-232 and RS-442) carried over cables to wireless communications using protocols like Bluetooth, Wi-Fi and GSM. Communications security (COMSEC) is the practice of preventing interception or access to communications traffic. Many of the COMSEC techniques are more relevant to military and government communications than to buildings. However, in addition to physical security, which was covered in the previous section, the major COMSEC disciplines relevant to design and implementation of building systems are:

(a) cryptographic security – protection of the communications traffic using encryption techniques to ensure the confidentiality and authenticity of each message. These techniques are useful to protect traffic that is being relayed over a wireless communications channel, for example, to prevent eavesdropping from outside the building and to secure connections between buildings in a campus environment. These techniques are extensively used in Virtual Private Networks (VPNs) to protect and encapsulate traffic flowing over an intermediary untrusted network.

(a) emission security – protection of the communications traffic from unauthorised interception and analysis based on the capture of emanations from the equipment. Computers, networks and communications systems typically radiate electromagnetic signals based on the information they are processing or displaying. In certain situations this information can be captured and analysed. This may result in the compromise of sensitive information that is being processed, transmitted over copper cables or displayed on a screen.

(b) traffic-flow security – protection to hide the flow of messages and message characteristics flowing over a communications network. Whilst this may not be a general issue in building systems, it may be of particular concern in electronic security systems, where an understanding of the traffic flow may reveal the detection and alarm capabilities of a system.

(c) transmission security – protection of transmissions to prevent interception and exploitation by means other than cryptanalysis, for example, the use of frequency hopping or spread-spectrum communication techniques. Transmission security also encompasses the protection of signals from jamming or interference, i.e. effectively radio frequency (RF) denial of service.

The need for the use of specific measures such as cryptographic, emission and traffic-flow security will vary from building to building and will be governed by the nature and severity of any vulnerabilities and threats. Common examples of where encryption is used in building systems are:

(a) in the set-up of Wi-Fi systems where Wi-Fi Protected Access (WPA) has been enabled to secure the communications between a computing device and the wireless hub. If WPA is not used then the network is effectively open, which can allow unauthorised devices to use it and/or access the wireless traffic; and

(b) in high-security buildings to protect alarm circuits by preventing interference going undetected and to provide authentication of the traffic.

G.3.2 Electromagnetic compatibility (EMC)

An area that is allied to transmission security and that is relevant to all building systems employing wireless components is electromagnetic compatibility (EMC), in particular, the need for adequate resilience to electromagnetic interference (EMI). All electronic, electrical and electro-mechanical technologies can suffer errors, malfunctions or failures due to electromagnetic disturbances, so it is necessary for effective measures to be taken to improve the EMI resilience of safety- and security-related systems. The increased use of wireless communications in fixed, handheld and mobile applications, both within and between systems, creates an increasingly unpredictable RF environment, which is getting progressively worse in terms of RF noise levels, spectral density and bandwidth.

Errors, malfunctions and failures due to EMI will typically be transitory, with little tangible evidence of their occurrence or the source of the interference. This makes fault identification, evidence gathering and analysis very difficult. The magnitude and distribution of potential sources of EMI is rarely known and, given the mobile or portable nature of many RF devices, the source of a problem may no longer be present when the investigation is underway. Within the European Union the Electromagnetic Compatibility (EMC) Directive [EC89/336 as amended] requires manufacturers and importers to satisfy specific requirements to ensure that the equipment they supply does not cause excessive interference. For many building systems, compliance with the EMC Directive should be sufficient protection.

For safety-critical systems and some security-critical systems, there may be a risk-based need for higher levels of EMC protection. The IET publishes a number of free Factfiles[10] that provide guidance on EMC as it relates to functional safety. These Factfiles are also relevant to situations where EMI could cause a malfunction in the cyber security of building systems.

G.3.3 Jamming

This is effectively a special case of an EMC problem, where the source of the EMI is the deliberate use of a jamming transmitter. Deliberate jamming is the RF equivalent of an internet denial of service attack. It has traditionally been a military capability typically involving the deployment of specialist personnel and systems to conduct electronic warfare in support of military operations. The cost and size of jamming equipment has fallen and this technology can be openly purchased online by the public. For those with modest technical skills, details can be found online describing how to build jammers and provide detailed designs. Deliberate jamming is no longer the specialist preserve of the military and is now available to anyone who chooses to buy or build a signal jammer.

To reduce the risk of inadvertent jamming of RF signals there are international agreements about the licensing and use of the radio spectrum. There are some licensing exemptions for short range devices (SRD): these offer a low risk of interference with radio services due to their low transmitted power. Examples of SRDs include remote locking transmitters for cars/doors, cordless phones, alarm and CCTV systems and radio frequency identification (RFID). In the European Union, with the exception of SRDs, all radio transmitters are required to comply with the Radio and Telecommunications Terminal Equipment (R&TTE) Directive [1995/5/EC as amended] and the EMC Directive referred to in Appendix G.3.2. Some unintentional jamming still occurs, particularly from the activities of illegal broadcasters.

A potential threat to building systems is the deliberate use of portable jammers, which can affect GNSS (GPS), mobile phones, Bluetooth™ and Wi-Fi signals. These jammers can affect an area from tens of metres to hundreds of metres in diameter. Whilst the use of these devices is illegal, they are relatively small, portable and easy to deploy. Depending on the frequency bands affected, these jammers could:

(a) disrupt the use of GPS signals as a source of timing information;
(b) interfere with the use of a GSM phone as an auto-dialler for alarms or to initiate remote support calls; and
(c) interfere with the use of Wi-Fi-based CCTV systems.

The probability and potential impact of a jamming incident should be considered for all building systems that use wireless components as part of their intra-system or inter-system communications. Depending on the impact there may be a need for specific countermeasures, such as the ability to detect the presence of signal

[10] *EMC for functional safety* – http://www.theiet.org/factfiles/emc/

jamming or an ability to automatically use non-wireless communications channels in the event of signal jamming or other EMI.

G.4 Systems architecture and interconnections

When faced with the need to provide cyber-security protection for building systems it is easy to rush off and implement obvious countermeasures, such as installing firewalls, deploying anti-malware software, etc. However, the indiscriminate deployment of countermeasures may not be the best use of scarce resources, finance or personnel and may create a false sense of security. It is considered good practice to start by fully understanding the risks faced by individual building systems and the building as a whole; this can be achieved by following the risk management guidance outlined earlier in this document. Once the risks are understood, pragmatic, appropriate and cost effective decisions can be taken about the countermeasures that should be deployed.

An important pre-cursor to the risk assessment process is the availability of information on the overall design of the building systems, including all components, locations, system dependencies and the reliance of the building and its users on individual systems. The understanding of overall building systems design should encompass the physical components, the software environment and the building data required or produced by the systems. The use of an appropriate enterprise architecture framework[11] can help to identify and document all relevant components. The enterprise architecture should encompass not only the technologies, but the human and process elements of the building systems as well.

As part of the risk assessment process, the interfaces to, and interconnections between, building systems should be critically examined. The interfaces represent an attack surface that may be exploited by threat agents.

Once the enterprise architecture for the building system has been established and the risk assessment undertaken, steps can be taken to reduce cyber-security risks and develop a secure architecture. This will encompass a variety of process, procedural and managerial countermeasures in addition to the technical measures discussed in this section.

G.4.1 Network and communications architecture

A common security problem with network and communications systems is a failure to keep track of connections, thereby losing control over an important part of the building systems' architecture. It is difficult to protect something where there may be unknown, unnecessary and unauthorised connections. It is important to address not only the wired (TCP/IP) connections, but also those using wireless technologies (for example, Wi-Fi, Bluetooth, GSM, LTE, etc.).

A first step is to identify all connections to the building systems. Once this baseline has been established, steps should be taken to minimise the number of connections to the building systems and to ensure that there is a valid business case for any remaining connections.

Wherever possible, networks used by building systems should be segregated or isolated from other networks, for example, corporate networks and visitor (guest)

[11] For example, the EACOE Enterprise Framework – http://www.eacoe.org.

networks with access to the internet. This is best achieved by implementing a dedicated infrastructure for mission- and safety-critical building systems. Wherever possible, remove any TCP/IP connections from the building's security and safety systems to other building's systems or networks. Where this is not possible, a risk analysis should be performed and appropriate countermeasures considered.

G.4.2 Firewalls

Where connections are required between building systems and other systems these connections need to be appropriately protected. The same principle applies when connections between building systems are required, where the systems have differing criticalities or security requirements. The standard IEC62443 (Industrial communication networks – Network and system security) introduces the concept of security zones and conduits to control access between zones. The standard defines a zone as a grouping of logical or physical assets that share common security requirements. It defines a conduit as a logical grouping together of communication channels, connecting two or more zones that share common security requirements.

Each zone will require a Security Level Target (SLT) based on factors such as equipment criticality and consequence of loss or failure. Equipment in a zone has a Security Level Capability (SLC) only if that capability is not equal to, or higher than, the SLT, in which case extra security measures, such as additional security technologies or policies, must be provided.

The conduits between zones should provide protection by resisting Denial of Service (DoS) attacks, preventing transfer of malware, shielding equipment in the zone and by protecting the integrity and confidentiality of network traffic. Any communications between zones must be via a conduit. The use of conduits can increase the SLC for all equipment in the connected zone.

Figure G.1 illustrates an IEC 62443 compliant approach, using zones and conduits to allow data from the building systems' local site computers to the business systems within the organisation (enterprise). In this illustration there are four security zones: a business or enterprise zone, a DMZ ('demilitarised zone') and two building systems zones. There are no direct connections between the two building systems zones or between these zones and the enterprise zone. Instead, firewall-protected conduits are provided, linking all three zones to the DMZ, with data or information exchanged between the zones via servers located in this central zone.

Figure G.1 – Security zones and conduits [adapted from ISO IEC 62443:2013, Figure A.2, page 70]

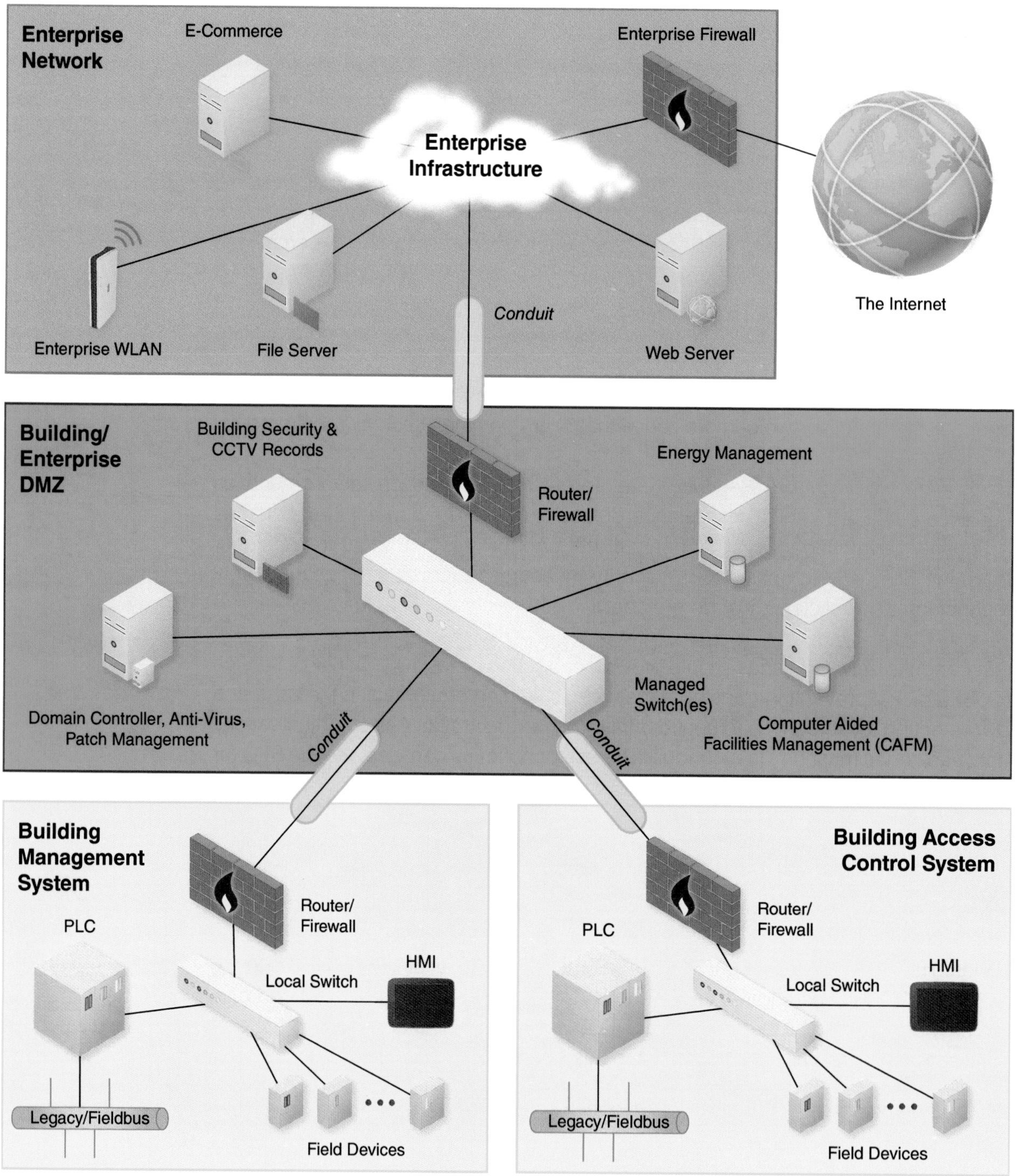

Where firewalls are deployed to protect the conduits, they should employ a tightly configured set of firewall rules. These rules bases should be appropriately documented so that the purpose and action of each rule is understood. The firewall configuration should be subject to regular review, with any changes managed under strict configuration control. The firewalls should be managed by appropriately trained systems administrators, who should implement an appropriate 24/7 firewall management and monitoring regime for each device.

G.4.3 Remote access

There are often requirements for building systems to have remote access connections, for example, to allow remote support and diagnostics to be provided by a supplier or manufacturer. These connections represent an attractive target for malicious threat agents as such connections often enable privileged access to critical systems.

An inventory of all remote access connections, their purpose and their type (for example, VPN or modem) should be maintained. The business justification for each remote connection should be reviewed regularly and steps taken to minimise the number of remote connections.

Where there is a suitable business justification for remote access, steps need to be taken to protect the connection. These should include the implementation of appropriate authentication mechanisms (for example, the use of strong authentication, wherever possible employing two-factor authentication). The remote access should be restricted to specific machines, for specific users and, if possible, at specific times and from specific locations.

Wherever possible, the remote access connections should only be made available when needed, for example, for a dial-up modem the connection to the telephone network could be disabled or unplugged when not required. Appropriate procedures and assurance mechanisms should be implemented to manage the enabling and disabling of remote access connections that are not permanently available.

It is important that basic cyber hygiene measures are in place for these remote connections, for example, the requirement that all remote access computers are appropriately secured with anti-malware, anti-spam and personal firewalls enabled. Regular security reviews should be undertaken of all third parties who have remote access to building systems. There should also be regular audits of access logs to ensure that any unauthorised access or anomalies are detected and investigated in a timely manner.

G.4.4 Anti-malware software

Anti-malware software should be installed to protect the building systems' workstations and servers. Appropriate mechanisms should be put into place to enable this software to be automatically maintained in an up-to-date state.

Where anti-malware software cannot be deployed, either due to the need to isolate systems or because the system vendor advised against installing the software on their systems, other protective measures should be implemented, for example, the use of gateway anti-malware scanning, manual media checking and system hardening (see Appendix G.4.6).

G.4.5 E-mail and internet access

E-mail and internet access should be disabled on all building systems that have security, safety or control functions. Where data is required by a building system, for example, weather data, this should be accessed via a demilitarized zone (DMZ) that is protected with appropriately configured firewalls and where the data is subject to anti-malware scanning before it is retrieved by the protected system(s).

G.4.6 System hardening

The hardening of building systems can be used to prevent or reduce the impact of network-based attacks by reducing the attack surface that is exposed. The first step in hardening a system is to remove or disable all unused services and ports in the operating system and applications, thus preventing unauthorised use. For the ports that remain open it is important to understand what services and protocols are used by devices connected to the network, especially for embedded devices (for example, PLCs and RTUs) that may use proprietary protocols. This may be established by using either a port scanner in a test or integration environment or the information that may be provided by the vendor. If a particular service is no longer required on a port it should be disabled. An example of a service that may be available but not used is an embedded web server.

Where a system has inbuilt security features these should all be enabled. Where systems are configured using default account names and passwords, the vendor's advice should be sought about the disabling of unwanted or unused accounts and the setting of systems-specific passwords. Where account names or passwords are changed the configuration should be recorded and stored in a secure off-line location that is only accessible to authorised users.

The use of removable media (for example, CDs, USB memory sticks or USB hard drives) should be discouraged and, if practicable, removable media should not be used at all. If this is possible the USB ports on the building systems should be disabled, either physically or by uninstalling/disabling the port driver software. If it is not possible, then procedures should be in place for the removable media to be checked for malware prior to use.

G.4.7 Backups and recovery

It is almost inevitable that at some point a building system will fail and need to be recovered. The impact of such a failure can be reduced by ensuring that there is a backup and recovery procedure in place that is effective for the identified electronic and physical threats. The backing up of systems software and configuration is the type of task that may be regularly postponed if it requires systems administrator intervention or initiation. Consideration should therefore be given to automating the backup process so that regular backups are made.

These procedures should be regularly tested as outlined in Appendix G.1. By testing the integrity of backups regularly through a full restore process, the systems operators will have confidence that the procedures work and any issues or uncertainties can be resolved without the pressure of a real business continuity event. It is important that any backups reflect the current configuration of the systems, i.e. new backups should be taken both prior to and immediately after any significant changes to the system, the former enabling fall-back, the latter ensuring an up-to-date copy is available.

Backups should be stored in both on- and off-site locations. Any media used to store the backups should be transported and stored securely, with appropriate measures employed to prevent physical or electronic damage to the backups.

G.4.8 Systems monitoring

Real-time monitoring of building systems can help to identify unusual behaviour that may indicate that a security breach or other electronic incident has occurred. For example, a sudden increase in the volume of network traffic may be an indication of a worm infection. To differentiate unusual behaviour from normal system operation, a number of parameters will need to be defined and monitored in real-time. For example, it is important to differentiate a surge in traffic volumes due to a change in a building operating state from an increase in traffic resulting from attempts to exfiltrate corporate data or malicious activity following installation of 'bot' software on a compromised machine.

Where practical, the implementation and use of intrusion detection systems (IDS) and intrusion prevention systems (IPS) within the building systems network environment may be used to provide a more detailed view of network activity. These systems will need to be tailored to a control systems environment and must not be allowed to interfere with normal system operation. Real-time monitoring of IDS and IPS tools can enable the early detection and response to intrusions, thereby limiting the access gained and any subsequent damage or harm.

The use of system logs to monitor system access and behaviour can be a useful tool, both to detect attempts to infiltrate the system and to provide material to support forensic investigations in the event of a cyber-security breach. A defined suite of system logs should be regularly reviewed and analysed. Tools are available to assist with the automated processing and analysis of system logs to identify unusual or anomalous records for operator analysis. Any anomalies should be investigated in a timely manner as post-breach analyses of many of the most serious breaches indicate that the attackers' presence was recorded in system logs often months before the serious damage was done. Unfortunately, in some instances the log files have been destroyed or badly corrupted by the attackers. It is therefore essential that important log files are backed up and protected against unauthorised access or modification.

In Appendix G.2 the physical security of building systems was examined. The installation of physical monitoring systems such as CCTV, tamper alarms on physical enclosures and intruder alarms for sensitive unoccupied areas can provide a warning of attempts to gain unauthorised access to building systems. The presence of these physical measures can also act as a deterrent, particularly in vulnerable or remote locations. The use and monitoring of physical access control logs can also be used to monitor unusual behaviour and identify potential insider threats.

G.4.9 Wireless networking and communications

In the built environment there is an increased use of wireless networking and communications. There are clear business benefits arising from the ease with which systems can be reconfigured and the savings accruing from the reduction in expensive fixed cabling. Unfortunately, wireless systems also introduce significant cyber-security risks and should therefore not be used in mission-, safety- or security-critical systems. Where the use of wireless networking or communication is proposed within a building system, a full risk assessment should be completed, addressing both operational and security risks.

Wireless solutions are constantly evolving, as are security features and threats. Solutions that are considered secure today could be regarded as vulnerable or obsolete within a couple of years. Where the risks associated with the installation of a wireless system are considered acceptable, it should be installed and secured in accordance with industry best practice. The use of the wireless systems should be regularly reviewed to ascertain whether industry best practice has moved on or if new threats render the continued use an unacceptable risk.

When designing and installing wireless solutions, care should be taken to understand the security mechanisms and ensure that they are correctly configured. Care should also be taken to review the siting of wireless access points to reduce the risk of interference or jamming and, as far as is practical, to restrict wireless access to locations within the secure perimeter of the building or site.

G.4.10 Security and maintenance patching

Most software requires regular patching to fix defects and address security vulnerabilities. To protect the building systems a process needs to be implemented to test and deploy security and maintenance patches. It is preferable for patches to be tested on a replicated/lab system or a systems integration system rather than being deployed straight into a live environment. The process should be supported by appropriate deployment and audit tools.

Where the patches relate to operating systems or other COTS software used in a building system, the process should allow for:

(a) vendor/supplier certification of patches;
(b) testing of patches prior to deployment – to ensure that it is a valid, authorised vendor/supplier patch that will work on the live system;
(c) staged deployment into the live environment, which will:
    i. minimise risk of disruption during the change; and
    ii. enable roll-back of the change in the event that problems occur.

Where patching is not possible or practical, the use of appropriate alternative control measures should be considered.

G.4.11 Device connection and disconnection procedures

Where a new device is to be connected to a building system or an existing device is to be reconnected, it is necessary to ensure that the device is free from malware infection. A procedure should be established to verify that it is not harbouring malware.

Where a device is removed from a building system there is a need to ensure that system configuration, PII and sensitive or classified data has been securely removed from the device. A procedure should be established to verify that devices have been successfully sanitised before they are removed from the controlled environment and/or the building.

G.4.12 Managing change

Changes to systems are inevitable, so appropriate change control and configuration management processes are required to support the operation and maintenance of building systems. Security assessments should be included in these processes. A proposed change should undergo formal assessment, which should include a consultation with all relevant stakeholders consulted. Depending on the nature of the change it may be necessary for it to be handled through both the building systems change management process and the building owner's or occupier's IT change management process, for example, changes that involve the connection of a building system to a corporate IT system or network.

G.5 Resilience

Resilience is the ability to rapidly adapt and respond to disruptions while maintaining continuity of business operations. From a business perspective, resilience is generally about preparing for any potential threat to the delivery of a smooth, steady and reliable service so as to maintain the delivery of critical services. From a building systems perspective, resilience is about:

(a) having fault tolerant systems, infrastructure and supporting processes that allow the building to continue to operate with minimum disruption in the event of the failure or a significant impairment of critical systems; and

(b) being able to recover, in a timely and efficient manner, from failure or serious impairment of critical business systems so that an acceptable level of functionality can be restored and the building can continue to be used.

To achieve this, the potential causes of disruption, both man-made and natural causes, will have been considered as part of the cyber-security risk management process described in Appendix B. Appropriate steps should have been taken as a result of this analysis to ensure that key building systems and their associated processes are maintained to deliver business continuity and that, where necessary, back-up systems and disaster recovery processes are available and rehearsed to enable timely detection of, and response to, disruptive events.

Whilst the concepts of business continuity and disaster recovery are reasonably well understood by organisations in respect of their corporate IT systems and their business processes, the complexity of the building systems and their dependencies may not be well understood by the organisation's business continuity planners. For business continuity purposes organisations that are heavily dependent on IT systems employ a range of provisions, including alternate/disaster recovery premises; off-site backups of business critical data; diverse network and communication routes, etc. These provisions will typically form part of an organisation's disaster recovery, incident response and business continuity plans. The nature of these plans and the specific measures required will be determined by the nature of the business, regulatory and legal requirements, and a business impact analysis.

The resilience of systems, whether they are IT or building systems (for example, HVAC) is generally measured in terms of redundancy, of systems and sub-systems, and their availability under both fault and maintenance conditions. Table G.1 illustrates a classification mechanism used for data centres and industrial plants. A building or plant classified as Tier 1 will have minimal resilience, with single points of failure in critical systems, and is likely to be used by organisations that can tolerate some loss of operation in their IT or building systems. A building or plant classified as Tier 4, however, will have a high degree of fault tolerance and may be used by an organisation that delivers critical national infrastructure services or that supports regulated financial and banking organisations. A Tier 4 site should be able to accommodate varying levels of scheduled maintenance and systems failure without losing capacity.

Table G.1 – Tier classifications for site infrastructure performance[12]

Tier	Description	Performance
1	Basic infrastructure	Non-redundant capacity components and single non-redundant connection/ distribution paths
2	Redundant capacity components infrastructure	Redundant capacity components and single non-redundant connection/ distribution paths
3	Concurrently maintainable infrastructure	Redundant capacity components and multiple distribution paths
4	Fault tolerant infrastructure	Fault tolerant architecture with redundant capacity systems and multiple distribution paths

[12] Adapted from – Site Infrastructure White Paper – Tier classifications define site infrastructure performance, by W. Pitt Turner IV, John H. Seader, Vince Renaud and Kenneth G. Brill. The Uptime Institute (2008).

How resilient a building's systems need to be would generally be determined by its operational use. For example, data centres and acute health care facilities will have requirements for the continuity of critical building services, whereas a retail outlet or warehouse may only require the provision of emergency lighting to allow safe evacuation of the premises. In these examples there may be varying levels of resilience within a building, for example, the computer halls and plant rooms may have resilient power supplies, whereas the power load in ancillary office or storage accommodation may be shed in the event of the loss of incoming site power. The analysis of systems' criticality and dependencies can be used to determine which business systems need to achieve specific levels of resilience.

APPENDIX H

Trustworthy software

Software is critical to the operation of building systems, whether it is the code in the programmable logic controllers that are in the cyber–physical systems, the human-machine interface (HMI) for the building management and access control systems, the databases used to store access control records or the applications used to create and manage BIM data. Few aspects of a building's operations are not touched in some way by a software-based system. This appendix provides information on the measures required to achieve software trustworthiness, and in the built environment these measures are potentially applicable to any software that is automated or has control function.

H.1 What is 'software trustworthiness'?

The concept of software trustworthiness can be defined through five characteristics:

(a) safety – an ability to operate without harmful states;
(b) reliability – an ability to deliver functionality or services as specified;
(c) availability – an ability to deliver functionality or services when requested;
(d) resilience – an ability to transform, renew and recover and respond in a timely manner to events; and
(e) security – an ability to remain protected against accidental or deliberate attacks.

These characteristics are consistent with the cyber-security attributes outlined in Appendix A.3. To help organisations understand how to develop and maintain trustworthy software, the TSI has developed the concept of Trustworthiness Levels and the Trustworthy Software Framework (TSF)[13], which acts as a consensus, providing a neutral way to access a range of existing good advice and standards. It is a layered repository with increasing levels of detail to service different user needs: the top level introduces methods and concepts, the second level addresses measures and principles and the third level encompasses specific techniques. This work has led to the development of a publicly available specification (PAS 754), which is available from the British Standards website[14].

H.2 Trustworthiness Levels

The TSF uses the concept of a Trustworthiness Level (TL) to inform decisions about the appropriate measures that may be required to ensure the stakeholders'

[13] References to the Trustworthiness Levels (TL) and the Trustworthy Software Framework (TSF) are licensed under the terms of the Open Government Licence v2.0. This is an open licence, which means that it allows information to be used and re-used with virtually no restrictions or charge. For further information see http://www.nationalarchives.gov.uk.

[14] http://shop.bsigroup.com.

needs are met. The TSF recommends the use of a 5-level scale[15], as illustrated in Table H.1.

Table H.1 – Illustration of trustworthiness levels

Trustworthiness level	Definition
TL0	Software trustworthiness not required
TL1	Software trustworthiness delivered in a due diligence manner
TL2	Software trustworthiness delivered by managed processes
TL3	Software trustworthiness delivered by established processes
TL4	Software trustworthiness delivered by predictable or optimising processes

Establishing the required trustworthiness level for an item of software is a four-step process:

(a) determine which facet(s) of trustworthiness are required, either explicitly or implicitly;

(b) assess the role that software plays in the overall system or service to be delivered, based on the degree to which the software is the source of trustworthiness:

   i. paramount role – the software provides the sole source of trustworthiness in the component, sub-system or system;

   ii. explicit role – the software provides the main source of trustworthiness in a component, sub-system or system;

   iii. implicit role – the software provides a major source of trustworthiness in a component, sub-system or system;

   iv. ancillary role – the software only provides a minor source of trustworthiness in a component, sub-system or system.

(c) assess the maximum impact that a defect or deviation in the software would have on the system or service. The potential impact can be assessed using a simple 4-level scale, where the impact is based upon the organisational context, for example: 'None', 'Routine', 'Significant' or 'Critical'.

(d) Using the TL matrix (see Table H.2), identify the relevant level role played by the software and its maximum impact.

Table H.2 – Trustworthiness level matrix

Role \ Impact	None	Routine	Significant	Critical
Paramount	N/A	TL3	TL4	TL4
Explicit	N/A	TL3	TL3	TL4
Implicit	N/A	TL2	TL3	TL3
Ancilliary	TL0	TL1	TL2	TL3

All assessments should be performed in a pragmatic, appropriate and cost-effective manner, using trustworthy software concepts, principles and techniques that suit the specific environment of the software and the building(s) it relates to. An

[15] This is an informative set of interpretations of the TL based on ISO/IEC 15504 (SPICE) concepts.

organisation should document its assessment of the required trustworthiness level for individual software components and use the results as a basis for prioritising efforts in software trustworthiness. The results from these assessments should be periodically reviewed throughout the software's lifecycle, to ensure that changes to the impact and role are regularly reassessed.

H.3 Trustworthy Software Framework (TSF)

H.3.1 TSF concepts

The TSF is based on four concepts:

(a) governance – the need to establish confidence in the trustworthiness of the software, which is achieved by having appropriate governance and management arrangements in place to address risk, control and compliance. What measures are appropriate will be determined by the stakeholders' needs and the environment in which the software is used.

(b) risk – an assessment of the risk that the software will fail to meet the users' and stakeholders' needs. This includes scoping those risks that are influenced by external dependencies, understanding the consequences of software failure, error or non-performance in view of the adversities (risks and hazards) that may be faced and ways in which the software may be susceptible.

(c) controls – software trustworthiness is achieved through the application of risk management, to enable identification and, where practical, the elimination of risks through the use of appropriate controls. These controls typically fall into one of the following categories:
   - i. personnel – related to the people involved in relating and using the software;
   - ii. physical – protecting the software artefacts and the operating environment;
   - iii. procedural – related to the specification, implementation and use of the software;
   - iv. technical – relating to the software environment itself.

(d) compliance – having adopted governance measures, understood the relevant risks and decided what control measures to adopt, developers and users of the software are required to implement the governance regime and apply the control measures.

H.3.2 TSF principles

The TSF assumes that when applying the four concepts in H.3.1, the organisation involved will follow the principles listed below. These principles are explained in more detail in Appendix H.4. In the case of a building system, the principles will apply to the building's stakeholders and the intended use of the building system and its software.

(a) *governance*
   - i. understand the general environment;
   - ii. understand the trust environment; and
   - iii. implement a formal management regime.

(b) *risk*
   - i. understand the general risks; and
   - ii. understand the trustworthiness risks.

(c) *controls*

i. personnel
- maintain practitioner competence;
- maintain organisational competence; and
- manage people risk.

ii. physical
- protect the physical environment; and
- provide artefact protection.

iii. procedural
- perform project management;
- perform supplier management;
- understand the requirements;
- maintain configuration management;
- confirmation of assurance;
- perform trustworthy software asset management; and
- maintain defect management.

iv. technical
- follow architecture-driven implementation;
- make appropriate tool choices;
- follow systematic design;
- seek trustworthy realisation;
- minimise risk exposure;
- practice hygienic coding;
- use methodological production;
- perform internal pre-release review;
- perform internal verification; and
- enable dependable deployment.

(d) *compliance*

i. performance acceptance verification; and

ii. maintain ongoing review.

H.4 Applying the Trustworthy Software Framework

H.4.1 Governance

Before procuring, producing or using any software that has a trustworthiness requirement, an appropriate set of governance and management measures shall be put in place. In a building, there will be a wide range of software embedded in components, sub-systems and systems. The levels of trustworthiness will vary from system to system and component to component depending on the software's role and the impact if it fails or does not perform as required.

GV.01 – Understand general environment

To establish the set of governance and management processes, the contextual background of the software and building should be understood. This will include the legal and regulatory environment, the technology to be employed and the nature of the building. It also needs to address the culture within the organisations responsible for operating and maintaining the building system, the building's owners/occupiers/users and the nature of any building systems-related extended supply chains. An approach to the assessment of the general environment was described in Appendix A.4.

GV.02 – Understand trust environment

To understand the need for software trustworthiness, the way in which the software will be used should be reviewed and recorded prior to acquisition or implementation.

This should include external dependencies, such as tenants, building users and customers, and the way in which these rely on or use the building. From this understanding, any special requirements (for example, for assurance, privacy and security) can be derived. An approach to the assessment of the trust environment was described in Appendix B.

GV.03 – Implement formal management regime

A formal management system should be instituted for building systems and should include specific accountability and responsibility in top management for trustworthiness of the building systems' software, along with the separation of functions, an explicit role of software release management and formal acceptance for third-party software, products and services.

H.4.2 Risk

Cyber-security related risk assessment processes were described in Appendix D.1 and include consideration of the set of benefits and assets to be protected, the nature of the adversities (threats and hazards) that may be faced, and the way in which the software may be susceptible to such adversities.

RI.01 – Understand general risks

In order to assess the risk to trustworthy software, an inventory of the software, building systems and building data is required. Taking account of this inventory:

(a) the totality of external factors likely to have deleterious effects on software should be identified. This superset is the aggregate of the set of hazards (undirected events) and threats (directed, deliberate, hostile acts); and
(b) the vulnerabilities of both bespoke and off-the-shelf elements should be understood such that an analysis can be completed, factoring in proportionality of protection.

RI.02 – Understand trustworthiness risks

In addition to the general risk assessment (RI.01), an understanding of factors particular to the specific trustworthiness should be obtained to include factors such as the maturity of technology, current and emergent weaknesses, attack patterns, malware and vulnerabilities.

H.4.3 Controls

In the context of software trustworthiness, risk should be managed by the application of controls. Where this is not possible, the risk should be managed through toleration, transfer or termination. A risk register should be maintained that records decisions about the treatment of the software-related risks. Controls that may be applied fall into the following categories:

(a) personnel – those measures applied to people involved with making trustworthy software.
(b) physical – those measures applied to protecting the trustworthy software artefacts and environments.
(c) procedural – those measures applied through the processes through which trustworthy software is specified, implemented and realised.
(d) technical – those measures achieved by the software environment itself.

H.4.3.1 Personnel

The 'people' aspects are examined in more detail in Appendix F.

PE.01 – Maintain practitioner competence

The various practitioners involved in the specification, realisation and operation of software shall be appropriately educated, trained and verified. This should include the use of continuing professional development (CPD) and the gaining of relevant experience, such that competence is achieved and maintained.

PE.02 – Maintain organisational competence

The organisation managing the building systems should have an appropriate management system in place. It should verify that appropriate competence levels are being achieved and maintained, both by its own personnel and by its supply chain.

PE.03 – Management of people risk

The organisation managing the building systems should have a single accountable owner of the people risks and ensure that people risks are comprehensively monitored. The issue of insider risk is addressed in Appendix F.

H.4.3.2 Physical

The physical protection of the building and its systems is addressed in Appendix G.

PH.01 – Protect physical environment

The environments in which software is specified, realised and operated shall be appropriately protected. There should be separation of development, test and operational facilities in accordance with the principles set out in ISO/IEC 27001. The physical protection of building systems is covered in Appendix G.2.

PH.02 – Provide artefact protection

Artefacts (including source code, software and systems documentation, and data) shall be protected from unauthorized access. For building systems that are required to achieve a trustworthiness level of TL3+, there is a need to consider the electromagnetic protection of platforms running the software and the cryptographic protection of sensitive data. The electronic protection of building systems is covered in Appendix G.3.

H.4.3.3 Procedural

These procedures relate to both the software and the platforms on which the software will run. They should be applied across the software lifecycle, i.e. to both the initial software baseline and to subsequent changes to it.

PR.01 – Perform project management

The project by which the building systems software is specified, implemented and realised should be planned, including the capture of requirements, the generation of explicit product descriptions and a process of peer review and validation implemented. This also applies to all changes to the software.

PR.02 – Perform supplier management

The supply chain for software used in the building systems should be understood so that trustworthiness can be specified and verified. Where the delivery of the software relies on the use of off-the-shelf libraries or packages, the trustworthiness of these should be understood and, for higher levels of trustworthiness, it should be verified. Aspects of supplier management are addressed in Appendices D.2 and F.2.

PR.03 – Understand requirements

The requirements for building systems software should be understood, including explicit (functional) requirements, implicit (non-functional) requirements and implicit, non-objective requirements. The user cases for software should be understood and any derived requirements arising during realisation should be recorded.

PR.04 – Maintain configuration management

All elements of building systems software should be subject to configuration management, including specification, realisation and release. This should include producing and maintaining documentation on the external constraints and dependencies involved in the building systems software deployment, configuration and operation. A formal product release and acceptance/commissioning process should be implemented, with release notices issued under the authority of the building systems release manager. Appendix E addresses cyber-security related configuration management issues.

PR.05 – Confirmation of assurance

To achieve confirmation of trustworthy software characteristics, an assurance case should be developed and maintained, which will form the basis for the assurance and acceptance review by the building systems release manager and the users.

PR.06 – Perform trusted software asset management

Processes should be implemented to manage the building systems software assets throughout their lifecycle. This should include the processes required for delivery, acceptance, asset recognition and review, and decommissioning. Appendix E addresses cyber-security related asset management issues.

PR.07 – Maintain defect management

All software defects identified both during realisation and in service should be recorded in a defect and deviation log, reported and assessed, with rectification at earliest opportunity using formal processes for the monitoring of deferrals. A triage process may be required for defects that affect complex building systems so that defect rectification may be prioritised based on the operational impact and the degree to which a defect causes any significant cyber-security vulnerability.

H.4.3.4 Technical

Appendix G.4 addresses specific cyber-security issues related to the architecture of building systems and the interconnections between systems.

TE.01 – Follow architecture-driven implementation

All software design should be based on an understanding of the architectural context, encompassing the properties of the building system in its environment as embodied in its elements and the inter-relationships of these elements, in accordance with the principles of ISO/IEC 42010. For building systems that are required to achieve a trustworthiness level of TL3+, consideration should be given to creating an architectural reference model from which architectural reference cases are produced, allowing appropriate generic design and/or effect classes to be selected, and architectural specification cases developed.

TE.02 – Make appropriate tool choices

Tools used throughout the building systems software realisation cycle should be appropriately selected to include development environment(s), programming language(s) and associated coding standards and testing tools. These tools should be configured such that their facilities that help enforce software trustworthiness are exploited.

Most programming tools, such as compilers, have options to check for weaknesses during code production, but these are frequently either ignored or switched off due to a perception that treating errors and warnings will slow down the software realisation process and require extra resources. Although this assumption may be true for the particular activity, many studies have indicated that whole-life time and resource expenditure is actually reduced by dealing with such errors and warnings at the first time they are encountered rather than retrospectively.

TE.03 – Follow systematic design

A system of formally developing and recording high-level design and low-level design information should be followed using, wherever possible, proven components (for example, libraries), and with specific procedures for the handling of third-party components (including open source software). The realisation of such designs should be documented.

It is important that, where software components from different sources are integrated as part of a system design, the assumptions informing, and the nature of, the design are understood. For example, where there is a data feed from one component to another, what does the receiving component do if that feed is interrupted? In a building system this could occur as a result of interference with a data feed from a wireless sensor.

TE.04 – Follow structured implementation

Bespoke components should be produced in accordance with good practice coding standards and the realisation of the implementation should be documented. This should include the use of relevant recognised data formats, algorithms, timing and synchronisation approaches.

TE.05 – Seek trustworthy realisation

When software is being designed, known failure and attack pattern modes should be reviewed, with components implemented in accordance with desired design/effect pattern(s), factoring in layered mechanisms to provide in-depth defence and, where necessary, appropriate cryptographic key management process and anti-tamper measures.

Within the building systems, mitigations for all identified failure modes should be implemented including, as appropriate:

(a) isolation for untrusted components (for example, sandboxing);
(b) isolation for high consequence code and data;
(c) malicious and mobile code control;
(d) control of network services and users, with appropriate access control mechanisms (authentication, authorisation and mediation);
(e) control of, and access to, log/audit/accounting/trace facilities; and
(f) provision of special controls (for example, passwords, cryptography) of appropriate strength.

TE.06 – Minimise risk exposure

To reduce the occasions when defects can arise or be exploited, only minimum privileges should be used, with all other actions defaulting to 'not permitted'. For example, a systems operator should not normally be monitoring a system when logged in with 'root' or 'systems administrator' level privileges. If these are required to perform a specific task, provision should be made in the design to allow the user to temporarily escalate their privileges, subject to the system's normal authentication processes.

All program data, executables, and configuration data should be separated. Software entry/exit points and use of interfaces to the system's environmental resources should be minimised.

TE.07 – Practice hygienic coding

To reduce the degree to which defects can arise or be exploited, coding approaches should be structured and aligned with coding standards. Examples of hygienic coding practice include:

(a) all variables, pointers and references shall be properly initialised at first and subsequent uses;
(b) all input data, messages and output data shall be validated;
(c) implementations of all algorithms shall be validated;
(d) error handling shall be comprehensive, fail safe and secure;
(e) a consistent naming convention shall be applied;
(f) resource access (for example, buffers, stacks, variables, macros, memory, cache and files) shall be explicitly managed; and
(g) detritus (for example, temporary files/logs) shall be removed.

For building systems that are required to achieve a trustworthiness level of TL3+, consideration should be given to including log/trace facilities that have the ability to audit the log/trace data.

TE.08 – Use methodological production

In order to understand the delivery approach for a particular implementation, before commencing build definition (i.e. the physical system configuration information and the various versions of software components and master data that form part of the system), customised checklists, integration standards, dependencies and assumptions should be produced and maintained. During the realisation of the building systems software, the organisation should enable and use compiler checking features, remove unused functions, configure components and perform unit testing before submitting components to integration.

TE.09 – Perform internal pre-release review

The internal integration and release function should perform:

(a) QA testing;
(b) load/performance testing;
(c) regression testing; and
(d) acceptance testing, culminating in the production of a software release note covering dependencies, assumptions and deferrals.

Depending on the required trustworthiness level of the software, the use of other techniques including exploratory, fuzz and penetration testing may be considered appropriate.

TE.10 – Perform internal verification

Verification of the building systems software should include:

(a) code analysis, including malware detection; and
(b) usability analysis, including consideration of the possibility of human error or misuse.

Depending on the building system, the required trustworthiness level and the assessed risk, other techniques that might be considered useful include:

(a) composition analysis;
(b) traceability analysis;
(c) fuzz testing; and
(d) penetration testing.

TE.11 – Enable dependable deployment

When the building system is being deployed a chain of custody for components should be maintained and the configuration should be made consistent with requirements, including granting only the minimum necessary privileges to system (software) processes and users. Once in use, the building systems software should be monitored for anomalous behaviour and a patching regime developed to allow for the application of routine, critical and emergency repairs.

For building systems that are required to achieve a trustworthiness level of TL3+, consideration should be given to executing code analysis and heuristic/behavioural monitoring of the implemented software.

H.4.4 Compliance

A compliance regime should be in place within the organisation that manages the building systems to ensure that governance, risk and control decisions have been implemented and are being maintained. Compliance regimes are necessary both within the organisation that produces the software and those using the software, i.e. the building systems' owner.

CM.01 – Perform acceptance verification

A process should be in place for accepting new building systems' software (including upgrades and patches) from the supplier. Verification of products/components should take the form of weakness testing, which might include fault injection. Verification of acceptance into service might take the form of penetration testing, fault injection and robustness testing. A subsequent and recurring compliance testing regime should be implemented.

For building systems that are required to achieve a trustworthiness level of TL3+, consideration should be given to employing independent specialists to perform the verification.

CM.02 – Maintain ongoing review

Once the building systems software is deployed, the management processes should be regularly reviewed to ensure that they are still relevant. Operational risk reviews should include checking progress against any deferrals in the software defects and deviations list. Internal audit processes should include reviews of any software issues encountered and metrics used, including efficacy of remediation of software defects and deviations deferrals.

The frequency of reviews will depend on the required trustworthiness level of the software. Indicative levels are:

(a) TL1 – yearly;
(b) TL2 – six monthly;
(c) TL3 – quarterly; and
(d) TL4 – monthly.

H.5 Application of TSF principles across systems lifecycle

The TSF principles set out in H.1 to H.4 spread out across a building system's lifecycle. Figure H.1 illustrates how the TSF principles can be mapped on to the systems lifecycle defined in ISO/IEC 15288:2008 [Systems and Software Engineering – System Life Cycle Processes]. The principles should only be applied to the element(s) of the system and software lifecycle as they are relevant to the organisation(s) responsible for the building systems.

Figure H.1 – TSF principles aligned to stages in ISO/IEC 15288 software lifecycle

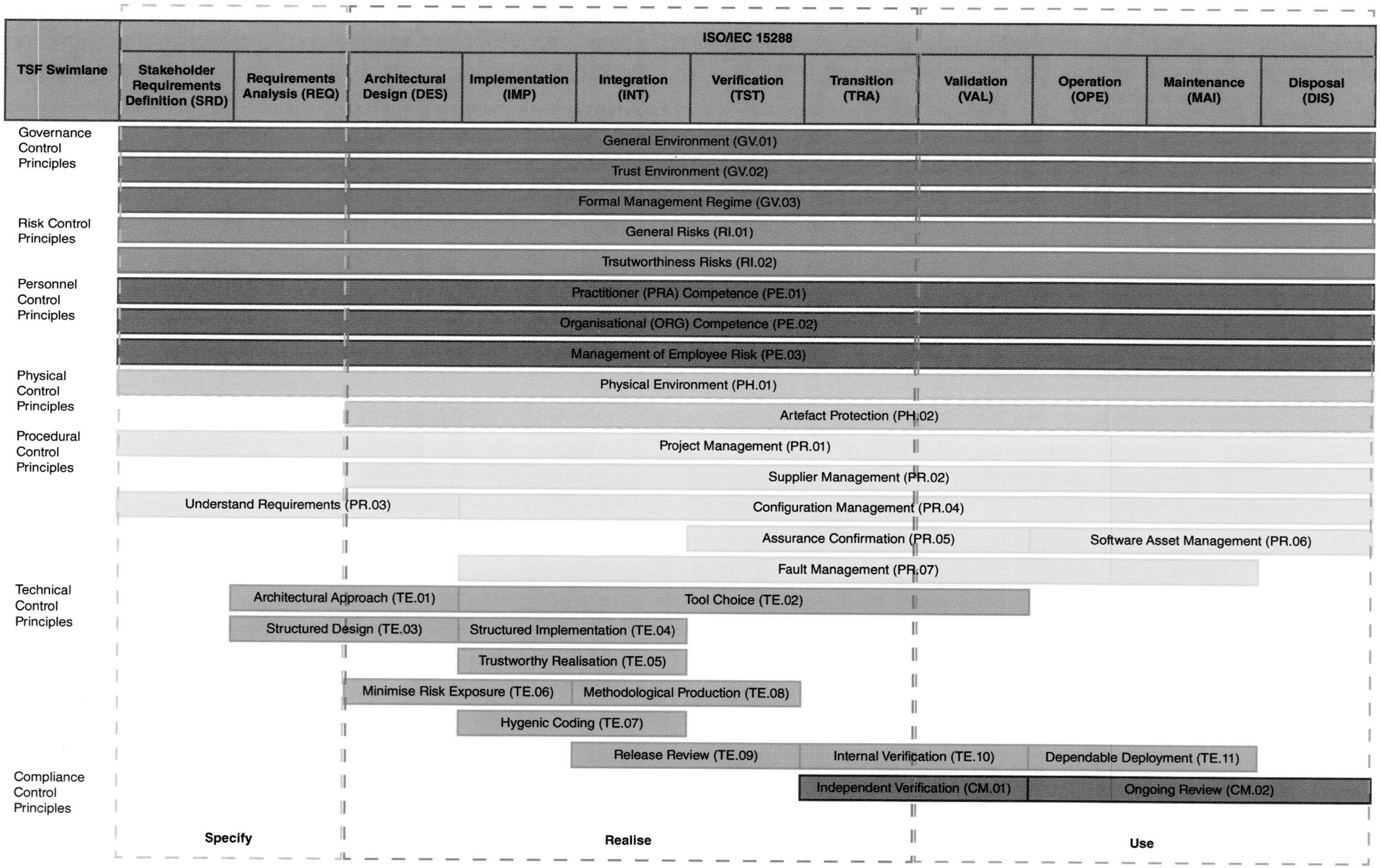

TSF Swimlane
ISO/IEC 15288
Stakeholder Requirements Definition (SRD)
Requirements Analysis (REQ)
Architectural Design (DES)
Implementation (IMP)
Integration (INT)
Verification (TST)
Transition (TRA)
Validation (VAL)
Operation (OPE)
Maintenance (MAI)
Disposal (DIS)

Governance Control Principles
General Environment (GV.01)
Trust Environment (GV.02)
Formal Management Regime (GV.03)

Risk Control Principles
General Risks (RI.01)
Trsutworthiness Risks (RI.02)

Personnel Control Principles
Practitioner (PRA) Competence (PE.01)
Organisational (ORG) Competence (PE.02)
Management of Employee Risk (PE.03)

Physical Control Principles
Physical Environment (PH.01)
Artefact Protection (PH.02)

Procedural Control Principles
Project Management (PR.01)
Supplier Management (PR.02)
Understand Requirements (PR.03)
Configuration Management (PR.04)
Assurance Confirmation (PR.05)
Software Asset Management (PR.06)
Fault Management (PR.07)

Technical Control Principles
Architectural Approach (TE.01)
Tool Choice (TE.02)
Structured Design (TE.03)
Structured Implementation (TE.04)
Trustworthy Realisation (TE.05)
Minimise Risk Exposure (TE.06)
Methodological Production (TE.08)
Hygenic Coding (TE.07)
Release Review (TE.09)
Internal Verification (TE.10)
Dependable Deployment (TE.11)

Compliance Control Principles
Independent Verification (CM.01)
Ongoing Review (CM.02)

Specify
Realise
Use

APPENDIX I

Bibliography

This appendix lists standards which are relevant to the design and operation of information and communications systems used in the management and operation of the built environment.

I.1 General IT security standards

Reference	Title/Description
ISO/IEC 13335	*IT Security Management – Information technology – Security techniques – Management of information and communications technology security*
ISO/IEC 15408	*Common Criteria for Information Technology Security Evaluation*
ISO/IEC 27001	*Information security management systems requirements*
ISO/IEC 27002	*A code of practice for information security management*
Critical Security Controls	*Critical Controls Version 5.0 – 27 February 2014* A reference set of recommendations for methods to address risks to enterprise data and systems. Published by the Council on Cyber Security (for further information, see http://www.counciloncybersecurity.org).
HMG IA Standard No. 1	*Technical Risk Assessment* IA Standard for Risk Managers and IA Practitioners responsible for identifying, assessing and treating the technical risks to ICT systems and services handling HMG information.
Supplier Information Assurance Assessment Framework and Guidance	Guidance on how the Supplier Information Assurance Tool (SIAT) question sets and tool specification can be used by suppliers of key business services to HMG.
Supplier Information Assurance Tool (SIAT) – Summary	A brief summary of the Supplier Information Assurance Tool (SIAT) Community of Interest set up to drive development of a supplier Information Assurance model. ISAB Approved.

Reference	Title/Description
CESG IA Top Tips	2010/01 – *DDoS – Distributed Denial of Service* 2010/02 – *Importing Data from External Networks* 2010/03 – *Basic Web Server Security* 2011/01 – *Trusted Platform Modules* 2011/02 – *Delivering Services Online* 2011/03 – *Mitigating Attacks to Online Services* 2012/01 – *Network Access Control*
BIS/12/1120	*10 Steps to Cyber Security: executive companion* Provides guidance for business on how to make their networks more resilient and protect key information assets against cyber threats.
BIS/12/1121	*10 Steps to Cyber Security: advice sheets* Provides detailed cyber-security information and advice on the 10 steps described in BIS/12/1120.

I.2 Security and safety of Industrial Control Systems (ICS & SCADA)

Reference	Title/Description
IEC 62443	*Security for Industrial Automation and Control Systems*
ANSI/ISA-99.00.01	*Part 1: Terminology, Concepts, and Models*
NIST IR 7176	*System Protection Profile – Industrial Control Systems – V1.0* Incorporates industrial control systems into Common Criteria
NIST SP 800–82	*Guide to Industrial Control Systems (ICS) Security*
IEC 61508	*Functional Safety of Electrical/Electronic/ Programmable Electronic Safety-related Systems*

I.3 Business-related security guidance

Reference	Title/Description
BIS/12/1119	*Cyber risk management: a Board level responsibility* Explains the benefits of cyber risk management to senior executives.
ISO 20000 / BS 15000	*IT Service Management Standards* Based on ITIL.
BS 7858	*Code of Practice for Security Screening of Individuals Employed in a Security Environment*

Reference	Title/Description
COBIT 5	*A Business Framework for the Governance and Management of Enterprise IT* (Control objectives for information and related technology.)
PAS 555: 2013	*Cyber-security risk. Governance and management. Specification*

I.4 Other standards and guidance

Reference	Title/Description
PCI DSS	*Payment Card Industry Data Security Standard*
NIST SP 800–61	*Computer Security Incident Handling Guide*
PAS 97: 2012	*A specification for mail screening and security*
RFC 2196	*Site Security Handbook* From IETF (The Internet Engineering Task Force)
RFC 2350	*Expectations for Computer Security Incident Response* From IETF (The Internet Engineering Task Force)
BS ISO/IEC 42010	*Systems and software engineering — Architecture description*
	EACOE Enterprise Framework

APPENDIX J

Factors to consider in assessing system context

This appendix relates the cyber-security attributes described in Appendix A.3 to the context described in Appendix A.4.

(a) *Confidentiality*

i. Human

- How do people interact (e.g. create, read, update, delete) with confidential building data across the lifecycle?
- What are the implications of these interactions?
- What interfaces do people use to interact with confidential building data?
- Who will need access to confidential building data?
- What access controls will be required to limit access to confidential building data?

ii. Information and data

- What building data may require protection?
- Where is the building data stored and processed?
- Does any of the building data require special protection from a privacy or personnel confidentiality perspective, for example, is it PII, such as health-related records for personnel?
- Does any of the building data require special protection in order to comply with specific legal or regulatory requirements and, if so, what are the requirements?

iii. Awareness and understanding

- What awareness do users have of the policy and procedural requirements related to the creation, use, maintenance, storage and transmission of confidential building data?
- Is there a programme in place to provide new users of the data with the necessary awareness training and for existing users with the delivery of periodic refresher training?
- Are maintenance and support personnel aware of their responsibilities related to the protection of confidential building data on systems they access?

iv. Electromagnetic spectrum

- What channels and technologies are used to communicate or exchange confidential building data between systems and people accessing the data?
- Are those channels confined to the building or do they allow the transmission of building data outside of the building?
- If the channels are not confined to the building or are accessible outside of the building, how is the confidential data protected?
- Do people access the building data remotely, for example, from other sites, from public places or from home? If so, how is this connectivity protected against unauthorised use and unauthorised access to the data in transit?

v. Building systems

- What systems are used to create, use, maintain, store and transmit confidential building data?
- Are these systems dedicated to a specific building or project? If not, what is the nature of the shared use and what measures are in place to protect the confidential data?
- When systems that process confidential building data are disposed of, what measures are in place to ensure that the data is securely removed from any storage devices?

vi. Infrastructure

- What infrastructure is used to communicate or exchange confidential building data between systems and the people accessing the data?
- Is that infrastructure confined to the building or do they allow the passage of building data outside of the building?
- If the infrastructure is not confined to the building or is accessible outside of the building, how is the confidential data protected?

vii. Environment

- What legislative or regulatory requirements affect the creation, use, maintenance, storage and transmission of confidential building data?
- Is the confidential building data only ever created, used, maintained, stored and transmitted within a single jurisdiction? If not, are there any differences in legislative or regulatory requirements in the jurisdictions involved?

(b) *Possession and control*

i. Human

- Who needs access to control/manage the individual building systems?
- What would happen if they were unable to control/manage the building systems?
- What could happen if an unauthorised user was able to control/manage individual building systems?
- Could loss of possession or control of individual building systems have health and safety consequences?
- Could loss of possession or control of individual building systems result in damage to the building or make it unusable?

ii. Information and data

- What information or data is needed to control/manage the individual building systems?
- Where is this information stored?
- Is it possible for unauthorised individuals to gain access to it?

iii. Awareness and understanding

- What awareness do users have of the policy and procedural requirements for the operation and management of building systems?
- Is there a programme in place to provide new users of the building data with the necessary awareness training and for existing users with the delivery of refresher training on a periodic basis?

- Are maintenance and support personnel aware of their responsibilities related to the protection of building systems from unauthorised access?
- Is there a programme in place to provide new maintenance and support personnel with the necessary awareness training, and for existing personnel with the delivery of refresher training on a periodic basis?

iv. Electromagnetic spectrum

- What channels and technologies are used to provide access to building systems for system management and control purposes?
- Are the channels used confined to the building or do they allow control access to building systems from outside of the building?
- If the channels are not confined to the building or are accessible outside of the building, how is this access protected?
- Would people need to access building systems remotely for control purposes, and if so, why? What would the consequences be if this access was no longer provided?

v. Building systems

- What building systems are in use?
- Are these building systems dedicated to a specific building or project? If not, what is the nature of any shared use?
- Where are the building systems controlled from?

vi. Infrastructure

- What infrastructure is used to manage and control the building systems?
- Is the infrastructure used confined to the building or does it extend outside of the building?
- If the infrastructure is not confined to the building or is accessible outside of the building, what protection is provided at the point of entry to the building?

vii. Environment

- What legislative or regulatory requirements affect the control and operation of the building systems?
- What legal or financial liabilities are associated with the control and operation of the building systems?
- In the event of a loss of control or possession of the building systems, who would be liable for any losses or damage?
- Are such losses insured or capped?

(c) *Integrity*

i. Human

- Are appropriate policies, processes and procedures in use to operate the building systems?
- Are appropriate policies, processes and procedures in use to maintain the building systems?
- Are appropriate policies, processes and procedures in place to manage changes to the building systems?
- Are appropriate policies, processes and procedures in use to maintain business continuity of the building systems?

ii. Information and data

- Is up-to-date configuration information maintained for the building systems?
- Are up-to-date backups maintained of all critical building information and data?
- Are appropriate policies, processes and procedures in use to maintain the integrity of building data?

iii. Awareness and understanding

- What awareness do users have of the policies, processes and procedures for the configuration management of building systems?
- What awareness do users have of the policies, processes and procedures for the operation and maintenance of the building systems?
- Is there a programme in place to provide users with the necessary awareness and training in the policies, processes and procedures, both for new users and refresher training for existing users on a periodic basis?

iv. Electromagnetic spectrum

- What electromagnetic interference could affect the integrity of the building systems?
- In the event of jamming or interference of building systems networking and communications channels, how would the performance and/or integrity of the building systems be affected?

v. Building systems

- Is the design of the building systems resilient?
- Are the building systems operated so as to maintain their integrity and coherence?

vi. Infrastructure

- What dependencies do the building systems have on infrastructure within the building?
- What dependencies do the building systems have on infrastructure outside the building?
- How do these dependencies affect or relate to the integrity of the building system?

vii. Environment

- What environmental factors could affect the integrity of the building systems?
- How can the impact of any of these factors be minimised or mitigated?

(d) *Authenticity*

i. Human

- Do users apply appropriate processes to the creation, reading, updating and deletion of building data?
- Do users apply appropriate procedures to ensure that only valid components (hardware and software) are used in the construction, configuration and operation of building systems?

ii. Information and data
  - Are appropriate quality criteria defined for the building data?
  - Are the information and data governance procedures defined?

iii. Awareness and understanding
  - Are users aware of the need to maintain building data quality?
  - Are users provided with appropriate training on the use of building data?

iv. Electromagnetic spectrum
  - How is authenticity of any communications or network traffic verified?
  - How would any interference with, or jamming of, communications or network traffic be detected?
  - In the event of an interference or jamming event, how are the building systems alerted to a potential loss of systems integrity or authenticity?

v. Building systems
  - How is the authenticity of system components (hardware and software) verified?
  - What measures are in place to test and verify the authenticity of components prior to their use in the building systems?
  - How is the authenticity of data created, processed and used by building systems verified?
  - In the event that invalid data is detected, what measures are taken to isolate the source of invalid data and prevent incorrect operation of the building systems?

vi. Infrastructure
  - How could the infrastructure be manipulated or altered to affect the authenticity of building systems or building data?
  - What measures could be taken to prevent or mitigate such manipulation or alteration?

vii. Environment
  - What environmental factors could affect the integrity of the building systems?
  - How can the impact of any of these factors be minimised or mitigated?

(e) *Availability*

i. Human
  - Can the building systems owner, building occupants and any support contractors implement the business continuity plans that will be used in the event of the failure or malfunction of building systems?
  - Are appropriate measures in place to prevent, detect and mitigate any hostile actions by insiders that may harm or disrupt building systems or lead to loss or corruption of building data?

ii. Information and data
  - Does the building's cyber-security policy include appropriate provision for the backup and recovery of essential building data?

- Do the processes and procedures cover the backup and recovery of all essential building data?
- Are backup copies of essential building data produced in accordance with the policy, processes and procedures?
- Does the building's cyber-security policy include appropriate provision for the creation and maintenance of a business continuity plan for the building systems?
- Do the processes and procedures cover the creation, maintenance and testing of the business continuity plans for the building systems?
- Are appropriate business continuity plans in place for the building systems?

iii. Awareness and understanding

- Are the building systems' owner, building occupants and any support contractors aware of the business continuity plans that will be used in the event of the failure or malfunction of building systems?
- Do the building systems' owner, building occupants and any support contractors periodically rehearse use of the business continuity plans that will be used in the event of the failure or malfunction of building systems?

iv. Electromagnetic spectrum

- Do the business continuity plans address the impact on building systems of loss or disruption of channels due to communications failures, for example, as a result of weather or solar events?
- Has the impact of jamming or RF interference on any wireless elements of the building systems been considered?
- How would the presence of a jamming or RF interference event be detected?
- Is the RF spectrum monitored in order to detect the presence of any jamming or interference?
- In the event of such jamming or RF interference, are appropriate measures available to maintain or restore affected building services?

v. Systems

- Have the building systems been specified, designed and implemented to meet the required availability?
- Are the building systems being operated and maintained to meet the required availability?
- Are the availability requirements adequately addressed in any support contracts for the building systems?

vi. Infrastructure

- Is the infrastructure within the building designed to meet the required level of availability?
- Is the infrastructure outside the building designed to meet the required level of availability?

vii. Environment

- What environmental factors could affect the availability of the building systems?
- How can the impact of any of these factors be minimised or mitigated?

(f) *Utility*

i. Human

- What user actions or loss of action could result in a loss of utility of building data or systems?
- What measures could be taken to mitigate the impact or prevent these actions or loss of actions?

ii. Information and data

- What measures are necessary to prevent the loss of utility of the building data?
- Are these measures already in place and, if not, what steps are required to implement them?

iii. Awareness and understanding

- Do the building systems' owner, users and any support contractors understand the need to main the utility of building systems and data?

iv. Electromagnetic spectrum

- Could interference or jamming cause a loss of utility of the building system or data?
- If so, what measures are necessary to prevent or mitigate the loss of utility?

v. Systems

- What measures are necessary to prevent the loss of utility of the building data?
- Are these measures already in place and, if not, what steps are required to implement them?

vi. Infrastructure

- Could damage to, or interference with, the infrastructure result in the loss of utility of the building systems or data?
- What measures could prevent this damage or interference?
- Are these measures already in place and, if not, what steps are required to implement them?

vii. Environment

- What environmental factors could affect the utility of the building systems or data?
- How can the impact of any of these factors be minimised or mitigated?

(g) *Safety*

i. Human

- What training, knowledge and experience do users require in order to operate the building systems in a safe manner?
- Do the users have appropriate knowledge and training to ensure that the building systems operate in a safe manner?

ii. Information and data

- What information and data do the building systems' owner, the occupants and any support contractors need in order to operate and maintain the building systems in a safe manner?

iii. Awareness and understanding

- Are appropriate education, training and awareness schemes in place to provide the building systems' owner, the occupants and any support contractors with the knowledge and experience they need in order to operate and maintain the building systems in a safe manner?

iv. Electromagnetic spectrum

- Could interference or jamming cause injury, loss of life or damage to the building due to a malfunction or failure of one or more building systems?
- If so, what measures are necessary to prevent or mitigate safety hazards?

v. Systems

- Are any of the building systems safety-critical?
- If so, do these systems meet the required legislative or regulatory standards?
- Could a threat agent cause these systems to fail to meet the required standards?
- What measures would be necessary to prevent or mitigate such threats?

vi. Infrastructure

- Are any special measures required to protect building and systems infrastructure from a safety perspective?
- Where equipment rooms are located within the building, are they equipped with suitable fire suppression systems?
- In the event of a fire or other event which damages the building, are critical building systems located so as to minimise or prevent damage from such events?

vii. Environment

- What regulations and/or legislation apply to the safety and use of the building?
- Are there any impending changes to these regulations or legislation?
- If so, how would these changes affect the continued operation of the building and its systems?

A

B

C

D

E

P